SUFFERING UP

MANASLU

David B Bethell

 FriesenPress

Suite 300 - 990 Fort St
Victoria, BC, V8V 3K2
Canada

www.friesenpress.com

ISBN
978-1-5255-6973-9 (Hardcover)
978-1-5255-6974-6 (Paperback)
978-1-5255-6975-3 (eBook)

1. SPORTS & RECREATION, MOUNTAINEERING

Distributed to the trade by The Ingram Book Company

To Dallas
Thank you so much for your generous support. I hope this book brings you inspiration.
Sincerely,

For my parents, Joyce and Samuel Bethell,
and my childhood friend, Elton

TABLE OF CONTENTS

Author's note.. vii

Preface.. ix

Part 1: The Canadian Rockies 1

Chapter 1: Elton's Story 3

Chapter 2: The Rescue 17

Chapter 3: Mt Robson, Canada's Great White Fright 23

Chapter 4: Himalayan Dreams... 35

Part 2: Journey to Nepal 39

Chapter 5: Into the city of gods, Kathmandu 41

Chapter 6: Monkey Business 49

Chapter 7: Highway of Tragedy 59

Chapter 8: Himalayan Vultures and Death Warmed Over 69

Chapter 9: Hear Tell of the Headless Goat...................... 83

Chapter 10: Basecamp and Beyond 93

Chapter 11: Himalayan Heatstroke.............................. 105

Chapter 12: My Heart-Wrenching Decision.................... 117

Chapter 13: Free Bird... 123

Part 3: Facing Mt Manaslu 133

Chapter 14: Suffering of the Sama and Happy Returns 135

Chapter 15: Hear Tell of the Headless Dog—Ruff, Ruff, Ruff! 145

Chapter 16: Back to Basecamp 157

Chapter 17: Mingma's Avalanche Rescue 167

Chapter 18: Manaslu Comes to Life . . . Earthquake!.............. 179

Chapter 19: My Furry Little Friend 187
Chapter 20: Ascension into the Abode of Shiva........................... 197
Chapter 21: This is Becoming a Bad Habit.................................. 209

Conclusion **219**

Glossary **221**

AUTHOR'S NOTE

Follow me as I take you on an incredible, dangerous and entertaining mountaineering journey, from the Canadian Rockies to the massive Himalayas in Nepal. In these pages I strive to gain the knowledge and experience to be able to climb Mt Everest. My mission is to raise awareness and funding to help children and youth who are slipping between the cracks of our social system and suffering in silence.

Between these covers you will be taken on travels and into tales that will bring tears to your eyes as well as great happiness to your soul. Your toes will curl in anticipation as I take your imagination away to the mountains bringing you into stories from the Rockies to the Himalayas. This non-fiction adventure story, written from the daily journals that I painstakingly kept, contains stories of conflict and heartbreak, but will also have you in stitches with laughter.

Truth is stranger than fiction and this book may well be one of the most entertaining compilations of non-fiction adventure that you will read. It will also give you a great sense of pride and tug at your heartstrings. With the purchase of these memoirs, you are making a real difference in bringing awareness and help to disadvantaged children and youth.

Dhan'yavad, (thank you in Nepali).

PREFACE

I have written this book in memory of my childhood friend, Elton, a disabled Indigenous child with a heart of gold. He took many beatings as he defended other children from neighbourhood bullies. Elton was an at risk child and only seven years old, and he suffered many atrocities in his short life. Sadly, I was often a witness.

Despite a tortured soul, Elton was sweet and he brought many smiles to me and other kids in our area. He had quite the sense of humour and loved making others laugh, even if he looked foolish doing so.

Tragically, little Elton suffered a severe injury in an accident in front of his home. He passed away just before the ambulance arrived.

Like my friend Elton, there are many at-risk children and youth who are suffering silently and feel abandoned as they slip between the cracks of our social system. With this book, I would like to bring Elton's story to light, as well as awareness of other children who are at risk. Elton's short life was not in vain, and will make a real difference for other children who are suffering in silence like he was.

In 2021, I will return to Nepal to climb Mt Everest under the banner of Canadian Rockies Youth Society. Through this climb, I'll be risking my life to raise awareness and funding to help at-risk youth, in memory of my friend Elton. As part of this journey, I also want to help the children of Samagaun, a village at the foot of Mt Manaslu where four in ten children die before the age of five. The name Manaslu means "mountain of the soul" or "spirit" in Sanskrit.

Proceeds from the sales of this book will be used to help pay for my expedition fees to climb Mt Everest but most of these I will raise myself, and I will donate the remainder of the proceeds to help charitable children/youth organizations, primarily in Alberta. If sales allow, I will also direct a portion of the proceeds to help the children in Samagaun. If not, I will use my own money; in my heart and soul, I feel I have adopted the village of Samagaun and its children.

I am resolved to succeed with my cause: it is dear to the core of my soul, as many of my school friends were at risk as well as me.

My parents, Joyce and Samuel Bethell, were generous people, fostering a few children before adopting me when I was two weeks old. My mom and dad did whatever they could to help my friends and other children in the area. My dad was ex-military and worked for Edmonton Telephones. He was also a part-time musician, photographer, and gave free music lessons to a couple of my friends whose parents couldn't afford lessons. My mother was a stay-at-home mom who had once managed a theatre and in her younger years played on a women's basketball team.

Today, my folks are deceased and I am without family of my own. At age fifty-five, I now realize my own mortality. I would like my mom and dad's hard work and sacrifices raising me to not be in vain. I essentially have one foot in the grave now, and I need to be able to leave this world with my soul in peace, knowing I made a real difference in the lives of children like Elton. I also want to leave this world knowing my parents would be proud of me.

Six years ago, I finally became mountaineer, realizing a childhood dream and finding a way to connect with the spirits of my dearly missed mom and dad. Since then, I have been increasing my skills and experience as a climber/mountaineer. I am now fully prepared to ascend the highest mountain of the world—Mt Everest—and I will use the opportunity to help at-risk children. This book is the story of my long journey towards Mt Everest, as well as my soul's ascension into heaven through compassion: all my experiences, misfortunes, heartbreaking stories, and close calls as well as successes while in the Canadian Rockies and the Himalayas.

To write this book (my first), I have drawn from the daily journals I kept while in Canada and Nepal. I know that you will find it entertaining; I myself have found that truth is also quite often more intriguing than fiction, and not only stranger. After this I plan to write a follow-up book about my experiences on Mt Everest and beyond.

My promise not only to you, my readers, but also to my friend Elton, my mom, my dad, and myself is this: I will strive until my dying breath to make a difference in the lives of children who are suffering and depressed, even if it only be a handful of kids. *Suffering Up Manaslu* and its proceeds will be a key factor in accomplishing this.

There is a real urgency to this cause: my attempt to climb Mt Everest is coming up quickly. By sharing my cause and compassion with you, I hope to manifest it into *our* cause—one that will benefit as many children as possible. In purchasing this book, I know that you will enjoy the stories in these pages as you feel compassion within your heart, knowing that you are a vital part in the solution for the children whose souls suffer alone and in silence inside the dark recesses of today's world.

Thank you for your compassion and support!

Sincerely,
David B Bethell

PART 1:
THE CANADIAN ROCKIES

CHAPTER 1:
ELTON'S STORY

From when I was a toddler to my teens, my family always took our vacations in the Canadian Rocky Mountains. We travelled widely in the mountains, but our favourite place was Jasper, Alberta, a quaint little mountain town with a rustic ambiance. My mom and dad's favourite spot there was Mt Edith Cavell, elevation 11,033 feet (3,363 metres). Until the early 1970s, on the foot trail was a teahouse we would visit, where we got peanuts that I would hand-feed to the chipmunks, squirrels, and various small birds.

My folks were great admirers of nurse Edith Louisa Cavell, who was stationed in a hospital in Belgium on the front lines of World War One. Most of the soldiers were young men in their late teens or early twenties, and Edith helped all who came to her, whether German or Allied. Because of her great compassion and lack of prejudice, Edith was nicknamed the Angel Nurse.

Edith also helped over two hundred of these young men escape back to freedom and their families. She wrapped them in bandages to hide them from the German troops in the hospital. While under the cover of dark, the young men would escape across the front lines to safety.

On October 12, 1915, the Germans had Edith executed by firing squad for helping these men escape, also accusing her of spying. One of the German soldiers on the execution squad put his rifle down, saying that Edith had saved the lives of his friends. He was punished severely for this, but refused to fire upon the Angel Nurse. As multiple bullets ripped through Edith's body and head she dropped to her knees while raising her arms up to meet heaven then collapsed on the ground.

Nurse Edith Louisa Cavell.   *Mt Edith Cavell with Mt Sorrow and Cavell Lake in the foreground.*

Mt Edith Cavell is absolutely beautiful (just like Miss Cavell herself). A glacier sweeps down out of the cirque in the shape of an angel with her wings spread as she ascends into heaven. This is so fitting for dear Edith, the Angel Nurse. On the other side of the cirque is Mt Sorrow, fittingly named for the sorrow felt after Edith's execution (which, by the way, was also one of the main reasons the United States became engaged in the Great War).

Every summer my family would go up to Mt Cavell, pay our respects, and enjoy the beautiful area—it is truly God's country up there. Off the bottom of the Angel Glaciers tail, waterfalls cascade down the rock into the ice at the base of Cavell Pond. Thanks to the glacial silt, the pond is a beautiful light emerald green. Even in the summer, glacial ice floats in the water like mini icebergs and rims the mountainside edge of the pond.

When I was five years old, my mother and sister were in the teahouse and my father and I were walking to the viewing area of the glacier. I heard a strange noise approaching us from behind: a jingling, clinking, clanging sound. I turned to see what it was, and coming upon us were two men with big packs, coiled ropes, and climbing gear hanging from them that jangled with their every step. As they came closer, my dad pulled me to the side of the trail to let them by. As they passed, they said hello and thanked my dad. They continued up the trail with ice axes in their hands and boots firmly planting their every step, with all the weight of the packs on their backs.

I was in total awe. These men seemed superhuman to me, and large with their big packs on. My dad told me they were mountaineers, explaining that they were going up to climb the mountain. Well, I felt like I just saw Superman—and not just one, but two of them! I could see and feel that my dad greatly admired these men.

This was the moment when something just clicked inside of me, and I knew what I wanted to do when I grew up. Every time I was in the mountains after this, I would be trying to climb little rocky areas or steep trails, pretending to be a mountaineer. At night, I would fall asleep to the smell and sound of the crackling fire in the stone hearth of the little rustic cabin, and dreamed of climbing Mt Edith Cavell. My passion earned the pride and admiration of my mom and dad.

At age six, I had a friend named Elton. He was seven and lived just up the alley. Elton's family was on social assistance and very poor. Both his mom and dad were severe alcoholics, spending what little money they had on alcohol. Elton and his younger brother and sister would steal food from the local stores so they could eat. Sometimes I would bring them to my house, where my mom would feed them.

Elton also had learning disabilities. Back then, he was diagnosed as being mentally retarded—although he wasn't stupid, just slower to learn. It took him longer to figure things out, but when he did, his memory was photographic. I soon understood if he was happy and felt safe, he opened up and could learn fairly quickly. He was a joy to be around with his quick wit and infectious laughter. He seemed to know full well that he was a little different from other kids. Even though, he didn't let this hold him back as he had a real zest for life. It was Elton who taught me how to ride a bicycle without using training wheels. He would let me use his old rickety bike to practice.

He attended a school called Winifred Stewart, which was for children who were slower to learn. Elton was the sweetest kid: very polite, would never harm a fly, and was always willing to help others with their yard work or chores. He didn't ask for anything in return, except maybe something to eat for himself and for his younger brother and sister. Sometimes he just wanted for the three of them to spend time away from the harsh conditions of their home. I know Elton and his siblings suffered many lashings from their father's leather belt in his drunken anger.

In my neighbourhood, we had many bullies. Whenever Elton saw another kid being bullied, he made it his business to stop it, even helping kids he didn't know. He would step in and confront the bully so the other kid could get away, and for this he took many beatings. On a couple occasions, he came to my mom to get patched up with band aids, and maybe just shown a little love.

I think that Elton was a sweet compassionate soul placed in the wrong body and wrong situation in this life. I now understand, though, that his younger sister and brother needed him in their lives. He was like mother and father to them, and did whatever he could to feed and fend for them.

One day when my mom was making dinner, I had the unnerving feeling that I needed to go see my friend Elton. My mom would not let me go because it was almost dinnertime. I wasn't hungry and had the desperate sense that something was very wrong with my friend.

Just after dinner, one of Elton's neighbours came over and told us the terrible news. Elton had gotten off the school bus as usual at about 4:00 pm, and as he was running up his sidewalk he tripped and hit his forehead on the concrete. According to his little sister Lisa, he went to his mom and dad for help, bleeding profusely from his forehead and crying in excruciating pain. This was a kid who was tough as nails, who had taken many beatings and never shed a tear. His parents, as usual, were heavily intoxicated and told him to leave them alone. Lisa said that he begged to see a doctor and they told him to go away. He then went to the bedroom he shared with six-year-old Lisa and his five-year-old brother, Mervin.

His older brother and sister were not home—they were in their teens and rarely home to begin with. Lisa tried to do what she could to help, and had the wherewithal to use a T-shirt for a bandage, but Elton's bleeding would not stop. At this point, it was about 4:30 pm. Elton told Lisa he felt tired and wanted to sleep. She left him and ran to the neighbours, frantically banging on doors, begging and crying for help. She was mostly ignored; this was in the early 1970s, and many people considered Indigenous people to be dirty savages, and wanted nothing to do with them.

When she finally got a response from the neighbours who were friends with our family, they called for an ambulance and went over to Elton's place to see what they could do. They were met with obstruction and anger from the intoxicated parents. The ambulance arrived in just ten minutes, but it was too late. Elton was gone. As he died, his sweet, tortured soul finally found some peace.

No one knew the whereabouts of his little brother Mervin until one of the ambulance attendants found him in the dark basement, hiding under the stairs, scared, in shock, and desperately trembling as he cried. After this, I believe social services finally stepped in and took the kids away. We heard nothing more as his mom and dad moved out of the house.

Because of Elton, his sister, brother, and other disadvantaged, depressed youth, I need to create a youth society. The goal is to raise awareness and funding to do what I can to help children with tortured souls who suffer in silence, enduring the hell they must survive in, or die in. This way, my dear friend's short life will not have been in vain, as I'm hoping to make a real difference in other children's lives in memory of Elton.

Last summer I went to the cemetery here in Edmonton where Elton's remains are resting, but I couldn't find a headstone or any kind of marker memorializing him. This doesn't surprise me, considering his family was very poor. One day soon, I would like to have a nice headstone placed for him, bringing to light the value of his short life and how his memory is helping other children through Canadian Rockies Youth Society.

As the years passed, we would continue to travel to the mountains every summer. I would try climbing more difficult rockfaces as I got older, but I always had an innate sense of when to stop in case I fell. My mom was constantly yelling out, "David, get down from there!" In my imagination, I was a mountaineer. It really is quite amazing how powerful our imaginations are as children! Sadly, many of us lose this gift as we get older and wrapped up in the rat race.

In my early and mid teens, we didn't travel to the mountains anymore. As I became more interested in girls, partying, and spending time with my friends, I lost my connection with the mountains, my mom and dad, and my boyhood dream.

By 2012, I was working at the Hinton Pulp Mill as a journeyman scaffolder. I was also suffering depression, mourning the end of a long-term relationship. My folks had passed some years before, and I had lost contact with my siblings (because I was adopted, they were much older and we were never close). I was on my own. But while working at the mill, I could see the mountains of Jasper in the distance, and felt them beckoning to me.

After finishing the job, rather than driving east to my home in Edmonton, I decided to drive west for a half-hour to visit the mountains. Once I entered the valleys between the peaks, an immense flush of emotion swept through my soul as so many memories came back to me. Deep down to the core of my every cell, I knew I was home. I felt I could sense the presence of my departed mom and dad. I spent that day blissfully visiting many places that hold memories of the happiest times in my life. The intoxicating smell of sweet mountain pine and wildflowers, the warm mountain breeze, and sunshine on my face took me back decades.

I then took a drive up all the switchbacks to Mt Edith Cavell. Well, this for sure was a step back in time for me! The teahouse was gone, torn down in the mid 1970s. I hiked up to the viewing area, stopping at the same spot where my dad and I let the climbers pass us. I looked up at Edith's mountain, and I'm not sure where it came from, but I knew then that I was meant to climb it.

That same day, I left the mountains for home with a renewed sense of worth in my life and a vow to come back the next year.

The following year, I went back and stayed at the Pine Bungalows Cabins on the banks of the Athabasca River in Jasper, where my family and I would always stay when visiting the mountains. The little rustic cabins have real stone fireplaces, and I managed to rent the same cabin my folks and I would often stay in. It hadn't changed in all these years; it even smelled the same with it's old woody scent. I went to the office and talked to the gentleman who now owns and manages the cabins. Michael is the son of the couple who used to run the place. They were good friends with my folks, and they always gave me toys whenever I visited as a boy.

Michael and I reminisced for at least an hour, and I told him of my plans to climb Mt Cavell. He told me the east ridge was too technical and dangerous, but there was a route around the west side where hikers can scramble up to the summit without climbing gear. He said it would be a long day, though, and some of the biggest grizzly bears in the area lived back there.

When I asked him if he knew the route, he pulled out a guidebook and photocopied me a map. Since this was June, he said it may be too early to make the summit as there could still be ice and snow. I said I could go do a recon trip and turn back if it became dangerous.

The following day I got up early. I drove up to Mt Cavell and parked my truck at the starting point. I then made my way to the top of the Mt Cavell-Mt Sorrow col (the shoulder where the two mountains join), which overlooks the cirque and Angel Glacier. Michael was correct there was too much ice and snow. Because I didn't have crampons, I turned around and went back down, but was happy rather than disappointed—the journey thus far was incredible.

Three months later, I came back. It had been a hot and dry summer, and I went to the local climbing shop in Jasper to inquire about the conditions on Cavell. They told me it was in great shape all the way to the summit via the west ridge route.

Saturday, September 14, 2013

I get my things ready, and at the last moment I decide to bring a headlamp, just in case. It's late in the season and the days are not as long. I park my truck at the trailhead and got on my mountain bike, heading along the trail in the Tonquin Valley around the back side of Mt Sorrow and Mt Cavell. (Mountain bikes are not allowed back there and the fine can be a couple of hundred dollars, but I decided the benefit outweighed the risk and continued on.)

Just before the main trail crosses the turbulent Verdant Creek, I spot the barely visible trail on the left that climbs into the dense forest. I push my bike about fifty feet up this trail, then lock it to a tree and out of sight. I hike up the trail talking loudly to myself to let any bears know I am coming. Back-country bears still have fear of, and try to avoid humans. Although you need to let them know you are there so you don't startle them or get between mama and her cubs. Using a loud voice works better than bear bells.

The forest here is amazing but dense, with old growth, young pine trees, and moss in shades of bright green. I can only describe this forest as being enchanted and a very spiritual place in the mountains. The earthy smell mixed with sweet mountain pine and alpine wildflowers is nothing short of transcendent and intoxicating to the soul. The young pine trees along the trail had soft needles that tickle nicely as they brush my skin. I need to make sure I keep speaking to myself out loud because if I come up on a bear, I won't see it in advance—the forest was that thick.

I finally emerge from the treeline into the higher alpine meadows and the cirque of the west side of Cavell. It's very hot for September. The weather forecast is for 34°C today. I look up and think that from either side, Mt Cavell is a massive mountain. I make my way up the scree and talus of the cirque, and then scramble up the rocky stepped sections to gain the Cavell-Sorrow col, which I had reached in June. After a snack and a break, I continue up the west ridge until I come to a steep shale section. I then traverse right to the southwest ridge—carefully, as a fall here would be bad! I follow the ridge to the summit.

I can't believe it! I make it to the summit of my first mountain ever, Mt Edith Cavell, which is also one of the "11,000ers" of the Canadian Rockies, at 11,033 feet (3,363 metres) in elevation. I'm so happy: out of any mountain, this is the most important one to me because of my family's connection to nurse Edith Cavell.

While on the summit, I have an epiphany. Wouldn't it be something to be able to help inner-city youth, like my departed friend Elton, who aren't able to get out of the city and away from their problems, bullies, or stressors? I could start a non-profit society to raise awareness and funding to help such kids. It could also benefit youth suffering from depression; getting out into nature has proven to be therapeutic, as I have experienced firsthand.

Looking towards Jasper from the summit of Mt Edith Cavell with Cavell Pond's green glacial water in the foreground and the crystal-clear blue Cavell Lake in the background.

As I take in the breathtaking views of both the Athabasca Valley and the Tonquin valley, I can feel the presence of my mom and dad's spirits. What a magical moment in my life! I can even see Mt Robson eighty kilometres off in the distance. I hang out there for about fifteen minutes. So cool: I'm now looking down on every other mountain in the area. I take a couple pictures and realize I have to begin my trip down—it's already 6:30 pm, as I had pushed the time a little just to gain the summit. I start back down, but carefully so I won't fall or get injured. I now notice the sun is dropping fast, and I'm concerned that I will not be down before the darkness swallows up the valley and then the mountain itself.

As I am coming down the scree slope, I slip, landing on my butt and ripping both cheeks out of my jeans. Great—now my ass cheeks are exposed and covered in scree dust! I'm okay, though, no cuts and just a few scrapes.

The light fading, I continue until I'm back at the high alpine meadow. *At least I have trail and good footing now*, I think. My next worry is that I still have about ten kilometres to hike out in dense forest, all alone, in heavy bear country, and in the pitch black. I can see the full moon starting to rise behind me as it chases the waning light of the day into utter blackness.

I enter the dense forest and start to remember all the stories my father would tell me about bears grabbing people in the mountains and dragging them into the woods. Some of the grizzlies in the Tonquin valley were reported to be eight or nine hundred pounds. I'm also getting worried about mountain lions, as sometimes they will not go the other way if they sense you. Instead, they will stalk you from behind and attack by pouncing on your back.

This is when I begin talking loudly, as though I have four or five friends with me, yelling out things like, "Come on Bob, catch up with me and bring the dogs!"

At 10:30 pm, I am deep in a charcoal-black forest, and the headlamp I brought isn't strong enough—I have to hold it down by my knee to see the trail. A couple times I walk off the barely visible trail, and the only way I can tell is when the ground becomes soft and spongy. I stop and back up until I am on the hardpack trail again, and then continue on.

Ahead of me, I see something on the trail. As I get closer, it becomes apparent that it's a big heap of bear scat. It appears fresh—the air is cool enough that I can actually see steam rising off of it. By the size and amount of it, I know that it was from a large grizzly (I've seen plenty of black bear scat, and it doesn't even compare).

Now I'm really worried. I figure I better keep moving. I had started ringing my ice axe on rocks along the trail, but now the ground is more moss, dirt, and tree roots than rock. When I am able to ring the axe on a rock, it sounds like a bell.

A few more feet down the trail, I can actually smell the bear. Grizzly bears have an unmistakeable, pungent smell. This is troubling, I'm downwind from the bear, the worst position I can possibly be in. It means the bear probably can't smell me or doesn't even know that I'm here.

I walk down the trail and ring my ice axe again. From a gully next to the trail, something big jumps up out of the brush and takes off behind me. As it leaves, it runs right over some young trees, one of which snaps back and hits me in the backpack, just missing my head as I turn to see what it was. I can just make out the outline of a huge grizzly, its shoulder hump backlit by the full moon as it takes off. My adrenaline is redlining—I think I might be leaving my own heap of scat on the trail in that moment!

After that, I make pretty quick work of getting back to my mountain bike; I have never been happier to see a bicycle! It is so nice to be off my feet finally and riding. There are a few shallow creeks that cross the main trail. They are easy to ride through, as they are only four inches deep. But I hit a big rock with my front tire as I cross one of these and somersault over the handlebars, landing flat on my backpack, looking straight up at the stars. I start laughing at myself, thinking how funny this would have looked to someone else. Good thing that my backpack cushioned my landing!

From the ground, I notice how beautiful all the stars look in the cloudless sky, remembering that Jasper is a dark-sky preserve for a good reason. I can clearly see the Milky Way painted across the sky, seeming so close that I could reach out and touch it. For a couple minutes, I just lie there and take in the view.

I finally get back to my truck about fifteen minutes later. What a relief— I'm now completely safe and have an incredible story to tell of the first mountain I climbed! Exhausted yet exhilarated, I drive to the Jasper townsite and set about finding something to eat.

I head towards the convenience store with a silly grin on my face as I think about what I have just accomplished. I pass a young couple and sense them staring at me . . . then hear them laughing. I glance over my shoulder and sure enough, they're watching me and still laughing. Then I remember that my ass is visible out of my ripped jeans. To top it all off, when I use the washroom in the store, I see that I am covered in dirt and scree dust from the mountain. I look as though I just crawled out of a garbage dumpster—no wonder they were laughing! With that grin on my face, I must have looked stoned or like a flashback from Woodstock.

* * *

After this—my very first ascent of a mountain, and also the mountain most dear to my heart—I decided that I wanted to fully realize my boyhood dream of becoming a full-fledged mountaineer/climber. I also decided to start a charitable youth organization, the Canadian Rockies Youth Society, in memory of my childhood friend Elton.

To gain the skills needed to be a climber and mountaineer, I began with a ten-week intensive course at rock-climbing gym Vertically Inclined. I followed this by training with Jay Mills, owner of Canadian Rockies Alpine Guides out of Canmore, Alberta. I started with his Glacial Travels and Crevasse Rescue course, then took his Alpine Skills Week, which involved climbing Mt Athabasca-2 (A-2) at 9,700 feet (2,957 metres) and Mt Cline, another mountain in the 11,000ers at 11,024 feet (3,360 metres). Mt Cline was also my first bivouac (or "bivy") on a mountain. I remember laying in my sleeping bag under the Milky Way, listening to rockfall all night long that sounded like thunder.

The following winter, I took Jay's Ice Skills intro course to begin technical ice climbing. I then completed his Ice Skills Week, when our group visited ice-climbing areas such as King Creek, Bear Spirit, and Haffner Creek. During this course, I discovered my favourite genre of climbing was mixed climbing and drytooling. Mixed climbing is when you are wearing crampons on your boots and you have the ice axes in your hands, but rather than just climbing ice, you are also climbing rock. With drytooling, you are only climbing rock with your crampons and ice axes.

I find these climbing disciplines to be most exciting because they are some of the scarier and more dangerous forms of climbing. Your crampons are popping off the rock. You have all these sharp tools on you, sometimes on the smallest of holds—it takes both upper body strength and finesse to pull yourself up. You must be powerful yet smooth, otherwise your tool can blow off the rock and hit you in the face or worse yet, impale you as you fall. The routes can be overhung, so sometimes you are actually crossing the roof of a cave by hanging from your axes one hand at a time, like a bat.

If you would like to see a great example of this on YouTube. Look up Sarah Hueniken's video, "Everyday," and at the end of this video you will see her on a route called Musashi. Sarah is now my mixed climbing coach and has been training me for this route which I will get back to training for and climbing once I have successfully climbed Mt Everest. Or check out "Sam Elias on Red Bull and Vodka, M11." (The M11 rating designates an expert mixed climbing route.)

During the Ice Skills Week with Jay, we were at King Creek in Kananaskis Country. I was on the vertical ice about sixty feet up, and I asked the guy who was belaying me to "Take!" This means he will take all of my weight on the rope to let me rest, as my arms were pumped and tired. As I was resting, he pulled his camera out of his pocket and dropped it into the creek. His reaction was to jump down and grab his camera from the shallow water, but in doing so he let go of the rope holding me up.

I fell—but luckily my reactions were good, and I managed to stop my fall by swinging my axes into the ice. Thankfully, they grabbed, as I had already built up some momentum. Once stopped, I had fallen eight to ten feet and had a sore shoulder, but after about an hour it felt okay. I think I just strained it a little. Thankfully I was able to stop the fall because it was sixty feet straight down.

For the rest of the course, I didn't let that character belay me again! Here's the ironic part—that guy is actually a pilot and flies a Lear jet for a living.

The next summer I completed a week-long advanced rock-climbing course with James Madden of Yamnuska Mountain Adventures in Canmore, joined by another good climber from Calgary. We spent most of the week camping and climbing in the beautiful Ghost Valley wilderness area. I enjoyed the course thoroughly, even though it finished with me at the Canmore Hospital for an extremely swollen right hand (I had an allergic reaction from a juniper needle poke between the knuckles when I reached in the bush to get my water bottle). The reaction was solved with Benadryl, after an hour's wait in the hospital.

After this I spent much time with Jay Mills, climbing routes such as the Gmoser route on Mt Louis, a mixed route on Loder Peak called Doors of Perception, as well as Mt Athabasca on June 3, 2016. Mt Athabasca is another 11,000er at 11,454 feet (3,492 metres), and for this climb my good friend Sherry joined us for her first full-on alpine ascent.

Sherry and I met the year before at the local championship drag races. She is from a small town outside of Edmonton, and a couple months after we met, she messaged me and asked if I knew any good places to hike. I took her to King Creek in Kananaskis Country and introduced her to ice climbing. This is the area where the movie, The Revenant with Leonardo DeCaprio was filmed.

A couple months later, Jay and I climbed Castle Mountain in Banff (elevation of 9,072 feet or 2,765 metres), ascending via the Brewer Buttress and Bass Buttress route. On July 25, 2016, we climbed Mt Edith Cavell via the world-class route up the east ridge, which was great fun. With stunning views, it's one of the most scenic routes in the world and requires a rope and anchors, unlike the west ridge. By this point, I also had done much more sport rock mixed and ice climbing in the mountains.

CHAPTER 2:

THE RESCUE

Six days after Jay and I climbed up the east ridge of Mt Cavell, I came back with my friend Sherry. Sherry had joined Jay and I on climbs before, including ice climbs, but this was her first time scaling Mt Cavell.

We would take the west ridge route. Being just a scrambling route at best, I figured this time I didn't need to bring a rope, harness, or any anchors, even though when I was on this route previously, I brought a rope and some gear out of habit.

Friday, September 20, 2013

It's great to be here and share the amazing forest of the Tonquin Valley with Sherry. We make great time until we get to the scree slopes leading up to the west ridge and the Cavell-Sorrow col.

I don't think Sherry is impressed with climbing the scree and talus. It's steep and loose, like climbing gravel, and is very tiring. We get onto the ridge directly above Angel Glacier, and then up to just below the summit, where you need to traverse a steep slope to the southwest ridge to gain the summit.

On the way up, we meet a couple climbers coming from the summit who ask me about the route down. I tell them to stay on the ridge until they get to the col between Cavell and Sorrow, because if they try to descend any sooner, they will end up getting cliffed out.

Sherry and I then reach the steep shale band under the summit where the traverse begins. Sherry is becoming irritable and I can tell she's tired. Just then, some fair-sized rocks fall, rocketing right across our path on the traverse. Because of this, I decide it's not safe to continue and we need to turn around and head for home.

On our way back down the ridge, Sherry is getting a little grumbly and complaining a lot. We're about two thirds of the way back to the Cavell-Sorrow col, and I'm tired myself. I can't wait to finish the climb and get back to my truck.

I notice a faint trail among the rock that looks like a shortcut to the trail down from the col. There's even a small cairn beside this trail, indicating that it's part of the route. We follow it and with some tricky down-climbing, we make our way onto a ledge about sixty feet above the scree slope.

It doesn't take me long to realize we aren't going to be able to get down. There's a spot to the left that might have been okay to down-climb if it were dry, but with water running over it, it's a no-go. The quartzite rock on Cavell becomes slimy and slippery when it's wet—treacherous to walk on, much less downclimb.

Sherry tries to convince me that we can get down this way, but she doesn't understand how greasy this rock becomes when wet. I spend about half an hour looking for a way down, but there isn't anything. We're cliffed out. So much for listening to the advice I had given to those other climbers!

The only way left is to climb back up to ridge and go to the col. But with Sherry being so tired and not having any experience with technical rock climbing, I don't think it's wise to have her climb back up the tricky section without a rope, harness, and anchor.

Figures, the one time that I don't bring a rope, I get us cliffed out and stuck! *I won't need the rope, we're just going to do a little scrambling,* I had said to myself that morning. We would already be rappelling down if I had the rope with me.

We still have about two hours until sunset. At least I was smart enough to bring my emergency locator beacon that sends an SOS signal to satellites for rescue. Our best option is to activate the beacon and wait for rescue, as we aren't equipped for a night on the mountain if the weather worsens. We try our cellphones, but there's no signal.

I pop the top off the beacon to let the bright copper antenna unfold. then push the button. Okay—it's on, and it also has a bright LED light flashing an SOS signal. I set the beacon on a rock away from the rock wall. Sherry and I then relax, drink water, and eat some of the food we had left over from the day.

The sun goes down a couple hours later. As it darkens, I wonder if the satellites aren't receiving the signal because we're in a cirque surrounded by the high rock walls of Mt Sorrow and Mt Cavell, as well as Chevron Peak directly to the west.

I then realize that no one is coming. It's too dark, and non-military helicopters are not allowed to fly in the mountains at night. We hear a jet fly over; it isn't an airliner, as it's lower and travelling faster. It sounds to me like a military or executive jet. I think that maybe they sent out a plane to get a location on us, which makes me optimistic. Maybe a military helicopter is about to be sent out—they can fly in the mountains at night thanks to instruments that show them where the mountains and terrain are.

There's no moon this night. It's pitch dark a few hours later, and the temperature has dropped with a cool breeze coming down from the snowy mountain. It's obvious now that we would be spending the rest of the night on the mountain, so we put on what clothing we had, place our backpacks flat on the ground, and sit on them with our backs against the rock wall. Sherry is exhausted so she lies down with her legs curled up on her pack and her upper body on my lap. She sleeps as I stay sitting upright with my back against the rock wall.

I feel a significant chill in my back—water has started to seep down the rock and is soaking my back. Sherry is sleeping so peacefully I don't want to disturb her, so I just stay with my back against the cold wet rock. The only thing keeping me warm now is Sherry sleeping on my lap.

I rest my arm on a small, shoulder-height ledge where my climbing helmet is sitting. I hear a light knocking, like my helmet is rocking back and forth from the breeze. This goes on for about two minutes, when I realize that the breeze had stopped.

Something runs across my hand, down my left arm, across my shoulders, and down my right arm, which is draped across Sherry to keep her in place as she sleeps. Startled, I fling my right arm up, sending whatever it is flying into the air and waking Sherry up. It must have been a packrat, which are common in the mountains. They like to chew on everything, including climbing helmets. I examine my helmet with my headlamp, and sure as the night is long, something has been chewing on the strap.

Sherry falls back to sleep, and I sit there watching the hypnotic strobing of the beacon flashing its SOS signal. It has been cloudy and I was concerned about bad weather coming in. But now the sky clears, making the wonders of the cosmos above us visible. I feel as though nurse Cavell is helping us out. Without the moon, the night is pitch black and perfect for viewing the night sky. I can see satellites and even the International Space Station passing overhead every ninety minutes, and the view I have of the constellations is breathtaking! This is one of those special moments in my life when I feel I am at one with nature. I'll never forget it.

I nod off a couple times, but only in short catnaps. I need to stay awake and alert in case any rocks begin falling. If that happens, I will have to wake Sherry so we can stand with our faces to the rock wall, hopefully out of the line of fire.

I am so happy to see the sky lighten up as dawn approaches. Sherry wakes up and I laugh. She had been drooling on my thighs as she slept, and I say jokingly, "So this is how I get a girl to drool all over me!"

The rising sun paints the peaks to the west of us in the warm orange and yellow shades of alpenglow. Thinking that the beacon had not been received, I decide I can safely leave Sherry here for a few hours now that it's light out. I will climb back up to the west ridge, where I know I can get a cell signal.

No sooner do I begin climbing, I hear a helicopter in the distance, and it sounds like it's coming our way. I climb back down to the ledge with Sherry and proclaim to her, "Our taxi is here!"

A couple minutes later, we see the helicopter fly around, searching for us across the valley near Chevron Peak. I pull out my bear banger flares and fire a couple into the air. They go off with a brilliant flash and a loud percussive bang! I notice the helicopter is an AS350 B3e, a type I have worked on many times. It flies over to our side of the valley and sweeps along the steep slopes, under the southwest ridge of Cavell.

Our rescuers finally spot us and land in the meadow below us. They hook up the 100-ft longline to the bottom of the helicopter, with the rescue technician attached to the end of the line. They fly up to us, and the rescue tech lands on our ledge then unhooks himself from the longline. The helicopter backs away from us to wait and hover at a safe distance from the rock.

He asks us how we're doing and puts a rescue harness on Sherry. We tell him that we are good. He calls the helicopter back, then hooks Sherry and himself to the line to be flown down to the meadow. After they transport Sherry, they fly back up to get me.

Once down in the meadow, they ask us if we had been moving around the valley during the night. They had received signals from three different locations in the valley. We tell them no, as we were stuck on the ledge all night. The cirque we were in must have bounced our signal off the rock around the valley, giving them erroneous locations on us.

Once they're all packed up, the crew fly us right back and land beside the parking lot on the other side of the mountain, where our trucks are parked. The rescue only cost us the entry fee into the national park, as this fee covers rescues.

Sherry and I are good friends to this day. We still go out ice climbing together occasionally, and now we have a great story to tell.

CHAPTER 3:
MT ROBSON, CANADA'S GREAT WHITE FRIGHT

On August 25, 2016, Jay Mills and I headed to Mt Robson (just under 13,000 feet or 3,954 metres elevation), eighty kilometres west of Jasper, Alberta. Mt Robson, also known as Canada's Great White Fright, is the highest mountain in the Canadian Rockies. Only about 10% of the climbers who attempt Mt Robson make it to the summit, often due to bad weather. The weather windows here are short, and the mountain makes its own weather that can change quickly and drastically. People sometimes say that Mt Robson belongs in the Himalayas: it's huge and takes up a massive amount of real estate.

Canada's Great White Fright, Mt Robson.

Mt Robson's prominence, or climbing elevation gain, is similar to Everest at around 11,000 feet (although Robson's oxygen levels are nowhere near as low as Everest). You start in the Fraser Valley at about 2,000 feet above sea level and then climb to just under 13,000 feet. This mountain gave me the bug to go to the Himalayas and climb Everest at 29,030 feet (8,848 metres) elevation. Before that, I would need to test myself and gain experience on another big mountain that was just above 8,000 metres, mostly to see how my physiology dealt with the altitude.

But first: Mt Robson.

Thursday, August 25/Friday, August 26, 2016

Jay and I arrive at the parking lot for Robson at 8:30 am. Jay fills out the paperwork at the notice board to let the park wardens know that we're climbing the mountain and when we expect to be back. Park wardens check the notice board daily, and start a search for anyone who hasn't returned by the date they specified.

By 9:00 am, we're trekking on the trail. Our packs are heavy: we're carrying all of our gear, food, tent cooking supplies, ropes, and clothing for any weather. My eighty-five litre pack was crammed full and heavy!

We trek past Kinney Lake as our route is to be up the Patterson Spur, and our weather looks good but only for two days. This means we are going to have to cover ground quickly. Just before Kinney Lake, the forest is amazing, like something out of Jurassic Park. I almost expect to see dinosaurs among the huge-trunked old-growth cedar trees that tower sixty to one hundred feet high. The plants and ferns here are also big, and seem like they are from the Jurassic age too.

Once we get past Kinney Lake and over a couple drainages, it opens up. We have to jump across a few small streams and then the fun begins. Up to 2016, this route wasn't used often, so it's hard to find the trail up a steep slope covered in young evergreens and juniper. We have a little bushwhacking to do. Before 2016, most climbing parties went around the west side of the mountain to the back and Berg Lake (named for the ice that calves off the Robson Glacier, forming ice bergs in the water). These groups then climb the Robson Glacier and onto the Kain Face. That route takes five to seven days. Our route cuts this time in half.

Eleven hours later, we make our basecamp just under a spot called "the knob." This spot was named for the prominent rock tower that looks like a knob. Once camp is made, Jay and I have dinner and prepack our smaller summit bags. We lie down in our sleeping bags for the night, but soon the sky unleashes on us. The torrential downpour on the tent sounds like we are under a waterfall. Unable to sleep, we just lie there until the rain finally stops around 11:30 pm.

We get up and unzip the tent to look out. It's beautiful out tonight: you can see all the stars and the Milky Way, and I think I can just make out the Crab Nebula. Okay, no sleep, boots, packs, headlamps on: it's time to climb. I look down behind us and the valley looks black except for the string of headlights from vehicles on the highway. They look like Christmas lights as they slowly meander their way up the Fraser Valley.

We climb past the knob and scramble up the technical rock sections. Here there are many big blocks of rock with deep voids between them. Then we climb over the ridge to the north side of the mountain in the dark. It's now 3:00 am and we are coming down the north side towards the Robson Glacier.

First, we must traverse a technical ice face to gain the glacier, using rope and ice screws for protection because we are exposed to falling thousands of feet.

We start up the Robson Glacier. It's riddled with crevasses that we must jump over or snake our way around cautiously. Jay and I are tied together with forty feet of rope, just in case. I am always ready to drop, dig in, and stop his fall should he punch through a snow bridge and drop into one of these bottomless voids! The headlamps in the dark actually make it easy to spot the weak snow bridges hiding the crevasses, as the low spots have a tell-tale shadow.

I follow Jay's footsteps in the snow, but about halfway up the glacier my right leg punches through. I drop up to my hip, feeling nothing but air below my foot.

"Guess what I found!" I yell to Jay while laughing as he looks back at me. I pull my leg out of the void, get up, and march on. Good thing it was a narrow crevasse.

Just before sunrise, Jay and I arrive at the base of the Kain Face (named after the famous mountaineer Conrad Kain, who first ascended Mt Robson in 1913). Above us is an endless ice and snow face at an 80° angle with about 1,000 feet in elevation gain. It's a calf-burner! I turn and look behind us to see the sky lightening as dawn approaches in shades of red and blue. Jay digs out a ledge in the snow to place our rope as it follows him up. I ask him if he wants me to belay him and he declines.

Jay climbs the face towards the end of the rope, about two hundred feet above me. "Fifteen feet!" I call, letting him know to stop with enough rope left to build an anchor. This is important—if I slip when climbing, Jay will use the belay and anchor to arrest my fall. "On belay," Jay yells, signalling he's secure and ready for me to start my ascent. "Climbing!" I respond. I begin to climb, slowly and surely until I set a nice rhythm with an ice axe in each hand. Right axe, step, step, Left axe, step, step—all the way up the two hundred feet towards Jay. The sun is coming up. Its orange and red rays reflect off the bottoms of the clouds, a glorious alpenglow on the Kain Face and Mt Robson itself. The colours are nothing short of spectacular.

The first section conquered, Jay and I continue the belay-anchor-climb sequence up the rest of the near vertical ice face. We reach the top of the face at 7:30 am. We can now rest for a few minutes and take in the spectacular views as my legs recover.

Next, we set the rope up for glacial travel with only about thirty feet between us, and then start climbing the east ridge of the southwest face. The snow up here is about two feet deep and the ridge is about 50°. We are not too far from the summit now. Some big seracs (ice blocks) overhang us and could fall at any time. It takes maybe another two hours to traverse the edge of a large crevasse, called a bergschrund (or schrund for short). Because of the angle, the other side of the schrund is about fifteen to twenty feet above us, with large feather-shaped icicles formed in the wind. I have never seen icicles shaped so strangely before.

The bottom of this bergschrund appears to be about two to three hundred feet down to our right. On our left side, the lip of the schrund we are walking on drops at a 70° angle down the south face into the valley below. If we both fall to the left, we would be on the grand tour down the south face but probably wouldn't survive. We also won't make it if we both fall into the schrund. Should I fall, I yell to Jay, "Falling!" and he would jump to the other side. The same if I saw him fall. Our fall arrested, we would climb back up to each other and continue on.

"Yahoo!" I yell out as we make the summit, at about 10:45 am. The weather is great, a light breeze from the west and some clouds in the sky. We take some video, pictures, eat a bit, and stay at the summit for about twenty minutes.

We notice the wind picking up and some dark clouds coming in from the west. We need to make our way home now, before any bad weather arrives. On our way down the east ridge back to the shoulder above the Kain Face, the wind strengthens and starts to gust. It feels like bad weather is headed our way. Something strange I experience is that I can't hear the wind until a cloud passes around us with a whoosh. Once the cloud has moved on, it becomes silent again, even though the wind is still blowing.

Jay above me ascending the Kain Face of Mt Robson, just before sunrise.

Alpenglow on the Kain Face of Mt Robson.

Ascending the southwest ridge of Mt Robson with Mt Resplendent behind me.

Jay and me on the summit of Mt Robson.

Down-climbing under the seracs.

The crevasse I found with my right leg earlier that day.

We are now at the top of the Kain Face and Jay builds a rappel anchor called a no-thread. For this, you use a long ice screw to bore two holes into the ice that meet each other. Then you take your rope and feed it through this borehole, pulling it out the other side until both ends of the rope are equal in length (you will run both sides through your rappel device). Now your rope is only half its length as you rappel on both halves of it. To alleviate this, you tie two ropes together, and you effectively have two hundred feet of rope to rappel on.

Before you rappel, the knot will be on one side of the boreholes. You need to remember which rope this is, usually by its colour (that's why it's best to have ropes of different colours). Before you throw both ends of the rope down to rappel, you also tie a stopper or safety knot in the end of each side, so that you do not accidentally rappel off the end of the rope and fall to your death. It's surprising how many climbers don't tie these safety knots, and sometimes die this way.

Jay rappels down first. Once at the end of the ropes, he anchors himself to the ice with an ice screw and detaches himself from the rappel ropes.

"Off rappel!" he yells to me.

I then clip myself into the rappel ropes and yell to Jay "rappelling."

Now rappel down to Jay and secure myself to the ice anchor he built. He then bores another no-thread into the ice, feeds the end of the green rope through the borehole, and ties a safety knot, preparing to repeat this sequence. First, though, we need to retrieve our ropes so we can reuse them.

The knot tying both ropes together was on the side of the green rope, not the blue one. We pull the slack from the green rope, stretch it out, and then jerk down on it hard. We keep pulling until it has fed through the anchor and freefalls down to us. Had we tried to pull the blue rope, we would have been trying to pull the knot through the borehole with no success.

When rappelling on this style of anchor, the pressure of the rope on the borehole will sometimes melt the ice, which will then refreeze, effectively welding your rope in the hole. That's why you must give the rope a good jerk to release it. On occasion, a rope welds itself so solidly that it will not release; in that case, the only choices are to climb back up to it or leave it behind. Better to get it if you can as ropes are not cheap!

Halfway down the Kain Face, the weather turns for the worse. Heavy clouds envelop us and snow begins to fall and blow around. As we set foot back on the glacier, the snowfall is heavy. Jay sets up the rope between us again for glacial travel, and we set off cautiously down the glacier, weaving our way between the bottomless crevasses and jump over the smaller ones.

Along the way, I see the crevasse that my leg found. It is only about a foot wide, but I can't see the bottom of it. In some spots, we can make out our footprints from the trip up, but can only follow these for a little while as they are swallowed up by the deeper snow. We make our way around a few crevasses that could swallow up large trucks.

The snow lightens up a little as we return to the ice face traverse. Jay puts in an ice screw to anchor the rope, and I belay him as he crosses. About fifty feet away, he stops and places another screw for protection and secures himself to it, marking where he will belay me from as I cross.

I start to cross and come to the ice screw. I sink my right axe into the ice, and kick my toe points into the ice. I then hang onto my left axe, which is also in the ice, and begin to remove the screw with my right hand leaving my right axe to hang.

Just as I pull the screw free, I lean in a bit, popping my toe points out of the ice. Ice screw in my right hand and axe in my left, I'm falling and rocketing down the ice face. I have the wherewithal to flip over onto my back and stick my legs out so I don't catch a crampon on the ice and break my leg or knee myself in the head. Because Jay and the belay anchor are about fifty feet to the horizontal of me, I go into a pendulum fall, and dropping very fast! Once I come to a stop, I'm at least sixty to seventy feet below Jay (the rope has stretched), but amazingly, I still have the screw and axe in my hands. Thank God Jay's anchors are solid. Wow, what an adrenaline surge! I scream out, "Wooo hooo!"

I now flip back over so I'm facing the ice and climb back up to the axe I left behind stuck in the ice. I retrieve the axe and get over to Jay and thanked him for the catch. Once we get off this face, back to the rock pinnacles, and over the Patterson Spur, the rain begins to come down in a deluge—so much so that the rock has become slimy and treacherous, as quartzite does when it's wet. Mt Robson and Mt Edith Cavell are similar in that they both have quartzite rock.

We briefly consider staying up here overnight, maybe heading back to the glacier to build a snow cave. We decide to continue on to our basecamp, keeping our crampons on for grip. The screeching scratching sound of the steel crampons on the rock is unnerving, almost like fingernails on a chalk-board. We also use the rope and go slowly—very slowly—because if one of us falls here among the blocks of rock, it would painful, if not bone-breaking.

Our plan works well. It takes time and we are soaked, but we make it back to the goat trail and rock steps of the spur, above the knob. Crampons off, rope put away, totally exhausted but almost to basecamp, thank God! We arrive at basecamp at 7:30 pm, get dry clothes on, have some dinner and crawl into the tent to finally sleep. We've been awake for about thirty-eight hours at this point. Total depletion is an understatement, as we've been on our feet climbing and descending in horrible conditions thirty-four of these hours.

Saturday, August 27, 2016

I sleep on and off most of the night. The rain on the tent is loud and keeps waking me up. By 8:30 am Saturday morning, the rain has stopped. Jay's awake and didn't sleep well, either. There is still cloud below us in the Fraser Valley, but it looks like it will burn off as the morning passes.

We break up camp and pack everything for the trip back to my truck. It's going to be a long day, especially with the full weight of the packs on our backs, carried by our tortured legs. (Jay is a machine, though; this is his second trip to the summit of Mt Robson in just over a week!) Before we finish packing, we eat some breakfast and bid adieu to basecamp.

We are now headed back down through the area where we rock-hopped over a couple creeks. Thanks to all the rain, these mild creeks are raging with heavy torrents. With some carefully placed footing, we cross successfully—until we get to the last and easiest creek. Jay is already waiting for me on the other bank when a rock rolls under my foot, and I slip.

I plunge into the rushing water headfirst, falling into a pocket where the current rolls me over onto my back. With my heavy pack on, I'm now as good as a turtle on its back. The cold water rushes over me, my arms and legs waving in the air as I try to flip back over. But it's not working!

Jay grabs my hand, pulls me out, and asks if I'm okay. "Good," I tell him as I laugh. He asks me if I'm cold. I tell him no. I was actually pretty warm before this happened, and I found the water to be quite refreshing. It cooled me down nicely.

It has turned out to be a beautiful sunny day. We follow the trail down over Kinney Lake and back through the Jurassic cedar forest, then back to the paved tourist trail beside the Robson River. The last four kilometres to my truck on this asphalt walking path seems never-ending, and is just brutal on the feet. To me, it is the hardest part of the whole trip.

Oh, what a feeling, my truck finally! How cool, Mt Robson completed—parking lot to parking lot in fifty-eight hours. Once back at home, all I did was sleep for a couple hours, wake up and eat something, sleep for a couple hours, get up and eat, sleep for a couple hours, eat again—nonstop like this for almost three days to recover. When I called and talked with Jay, he had also recovered from his two climbs of Robson pretty much the same way: eat, sleep, eat, sleep, repeat.

CHAPTER 4:

HIMALAYAN DREAMS

After recovering from Mt Robson, I began to wonder how I was going to help youth through the Canadian Rockies Youth Society. Then it dawned upon me: I could use climbing as a way to raise awareness and funding. But it was going to take climbing a notoriously dangerous mountain—the biggest in the world, Mt Everest.

Up to this point, I had no interest in climbing Everest, or even any mountains in the Himalayas. I prefer technical climbing, and from what I've heard, these big mountains aren't what I would consider real climbing—more like hiking to me. I never really did get into hiking, only doing so to get to a climbing area or into the back country to climb a mountain. (Climbers call this "approaching the climb," not "hiking.")

I did some research on the Himalayas, and learned more about K2 and Mt Ama Dablam, as well as others mountains like Meru Peak. These piqued my interest: they were very technical, but not as well-known as Everest.

I decided that I was going to need to put my technical, scary, and fun climbing on the back burner, and focus first on helping children and youth like my friend Elton. This would also make my mom and dad most proud of me! I knew, though, that I couldn't just fly to Nepal and walk up Everest.

The mountain has taken many lives due to altitude, weather, falls, the cold, exhaustion, avalanches, and falling seracs, not to mention overconfidence and stupidity, also known as summit fever.

I smartened up, realizing this style of climbing was a completely different animal, one that doesn't bare its teeth until it is too late. I couldn't take this lightly.

I contacted Lakpa Thendu Sherpa of Adventure 14 Peaks in Kathmandu, and told him of my plans. His advice was to first climb a smaller peak over 8,000-metres, such as Mt Manaslu. After talking with Lakpa, I started researching Mt Manaslu and other people's experiences on this huge mountain.

Mt Manaslu, mountain of the soul.

After a few days, I called Lakpa back, wanting to be part of the Autumn 2017 Manaslu Expedition. Lakpa tells me it is $11,500 USD for the expedition, including full board service. It's the same as all-inclusive, excluding the flights to Nepal, personal spending, and climbing gear. Lakpa sent me the itinerary and the list of what I needed for this climb. I already had most items, and the rest I could purchase or rent when I get there.

At the time, I was working nights north of Fort McMurray, Alberta, and it was cold. One night, I was part of a gear chain, passing materials up the scaffold to the guys building the scaffold at the top. Tonight, the temperature was -36°C, -44°C with the windchill. Waiting on the scaffold deck, I paced back and forth to stay warm—but in doing this, I strained the meniscus in my right knee.

I went to medical on site, and they put me on light duty in the office doing paperwork. On my days off, I saw my physiotherapist, Andreanna, who has performed miracles with my past strains and injuries. She works on my knee, shows me some exercises that help, and tells me to ride my bicycle daily.

By spring, I was no longer at that job and was working day shift. At home and on my days off, I was out on my mountain bike every day to rehab my knee and get in shape for Manaslu. Occasionally I would take trips up to Jasper so I could trek and do some scrambling at elevations over 10,000 feet for high-altitude training.

On July 2, Jay and I climbed a rock route called Econoline, on the east end of Mt Rundle near Canmore. I told Jay about my knee. At first we were a little concerned, but it gave me no trouble and we have a great day in the sunshine at 24°C.

That was two months before leaving for Nepal. I continued my training in Jasper as well as mountain biking and weight lifting right up until a few days before I leave.

Tuesday, August 29, 2017

Today's the day I leave for Nepal. First thing in the morning, I drive across town to my mom, dad, and grandma's resting place, the beautiful Mount Pleasant Cemetery. It's a park-like setting, with hills and trees. The smell of pine reminds me of time spent as a boy with my folks in the mountains.

The morning at the cemetery is warm and sunny. There are always many finches, chickadees, and squirrels here, as well as the occasional rabbit. I sit in the grass, take in the sunshine, and chat with my mom and dad, thinking this could be the last time I visit them here if things go bad on Manaslu. Before I leave, I ask my folks if they can come with me in spirit and experience the beautiful Himalayas through my eyes, as I know they both would have loved this trip immensely.

I go back home, park my truck in the garage, and finish up last-minute details like pre-ordering a taxi to take me to the airport. It begins to hit me: I am really travelling halfway around the world to the Himalayas, how cool is that?

My taxi shows up and I'm off. I fly from Edmonton to Vancouver, Lay over for a few hours then take a twelve-hour flight that goes up the west coast, then over Alaska and the Bering Sea, where I can actually see crab-fishing boats in the water below. We then head west over the international dateline and follow the Russian coast along the Sea of Okhotsk.

Next, we fly over Japan and South Korea. This is the same year that North Korea is conducting failed rocket tests, with pieces reported to have fallen into the Sea of Japan, right under our flight path. *I sure hope Kim Jong-Un doesn't conduct any tests now!* I think as I look out the window.

We land at Hong Kong's massive airport. I think it's more a fly-in shopping mall, as it's full of many high-end stores like Rolex, Gucci, Versace, and more. The smog here is bad; I can't even see across the harbour, much less the airport. What a busy airport this is, and it's so cool to see the new Airbus A-380 double-deckers coming and going. These new airplanes are so massive, they make a 747 Jumbo Jet look tiny. The A-380s are impressive to watch, as they appear to take off in slow motion due to their massive size.

A few hours later, I'm on my way to Kathmandu, a four-hour flight from Hong Kong. With layovers, it has taken almost two days to get this far. I haven't slept much since leaving home, and I'm exhausted.

About twenty minutes away from Kathmandu, the weather turns nasty, with electrical storms, heavy rain, and turbulence. The light show outside my window is incredible as the flashes of lightning backlight the massive storm clouds. The airplane was bouncing and shuddering as we were experiencing heavy turbulence I remember from the TV show Mayday that Kathmandu's Tribhuvan Airport is one of the most dangerous airports to land at, because it's in the Kathmandu Valley surrounded by steep hills and the massive Himalaya mountains.

Interestingly enough, once we settle into our final approach into the Kathmandu Valley, it's like someone flipped a switch. The lightning and turbulence stops as we leave the thunderstorms behind.

Awesome, I think.

PART 2:
JOURNEY TO NEPAL

CHAPTER 5:
INTO THE CITY OF GODS, KATHMANDU

Wednesday, August 30, 2017

We are now about five minutes from landing, and they are playing nice relaxing music over the speakers. *Great,* I think, *they are playing music to make the crash pleasant and enjoyable.*

As we come in over the ancient city of Kathmandu, I can see dark narrow streets and buildings lit up with multicoloured lights. The plane touches down at 10:30 pm local time. When I disembark the plane and walk down the stairs, the heat and humidity hits me and it feels like I just walked into a steam room. I can also hear the sound of crickets. Walking towards the arrivals door, the crickets become louder as they appear to be in some tall grass next to the building, their night song soothing to me in my state of exhaustion.

In the terminal, it's sweltering, with no noticeable air conditioning. As we line up to fill out our paperwork and get our visas. I notice that even the locals are sweating the same as me. So it's not just me, it's indeed very hot in here!

It only takes about fifteen minutes to get through, pick up my bags, and get outside, where I immediately recognize Lakpa Sherpa, the managing director of Adventure 14 Peaks. With Lakpa is the driver of the hotel where I will be staying. As I come up to them, pushing my cart full of bags, Lakpa places a wreath of marigold flowers over my neck and a scarf with Tibetan prayers printed on it. He welcomes me with a big smile and "Namaste." With myself and luggage loaded into the SUV, we make our way through the busy streets of Kathmandu.

Along the way, I saw many stray dogs and felt sorry for them. What I don't understand yet is that all the stray dogs are well taken care of here. It's integral to the Hindu and Buddhist cultures to respect and care for animals, as they carry souls like we do.

I also see many buildings still being repaired from the damage caused by the 7.8-magnitude earthquake two years earlier. The 2015 disaster took out the basecamp on Mt Everest, killing nearly 9,000 people and injuring 22,000 throughout Nepal.

Many of the streets here are narrow, without sidewalks. They have just enough room for one vehicle. After about ten to fifteen minutes, we arrive at The Sacred Valley Home Hotel, my home while here in Kathmandu. The hotel is in the Thamel district, the tourist hub of Kathmandu.

It's now almost 11:15 pm local time, and around 11:00 am at home in Edmonton. I feel like a zombie—exhausted not only from lack of sleep, but also the drastic time change. I'm greeted by the hotel manager (Nabin Giri), his son (Aashish) and his nephew (Samir). All have welcoming smiles and help get my bags up the marble stairs to my room on the third floor.

My room is spotlessly clean, with a full washroom, shower, double bed, TV, desk, air conditioner (I immediately turn it on to cool the room), and a balcony with a nice view down the street. The walls, unlike home, are all made of solid concrete, but then again we don't have earthquakes at home.

Okay, time to curl up in this nice bed, and finally get some sleep. *"Subha ratri"* ("Goodnight" in Nepali.)

Thursday, August 31, 2017

It's 6:00 am and I think I only got about four hours sleep. Just after 3:00 am, there was a hell of a boom. It was so loud it shook the building and scared me out of my sleep. I literally jumped up and out of bed. I looked out my window to see that the lightning storms from my flight had followed me to

Kathmandu. Lightning had hit a power pole across the street from the hotel. A few wires on this pole were showering sparks and still smouldering. It had also caused the power to go out, as my air conditioner was now off. After a few minutes, my heart rate came down, and I laid on the bed and fell back to sleep.

This morning, I get up to the sound of roosters crowing, which I did not expect in a city of this size (Kathmandu's listed population is 2.5 million). I get up and have a nice steaming-hot shower. After getting dressed, I go upstairs to the rooftop café to have breakfast with Lakpa Sherpa.

My mom would have loved it here: an open café on the roof, tables with colourful umbrellas, flowering plants, and even a couple trees in large clay pots. Besides the plants, there are also small birds chirping and visiting all the planters for food, and bees buzzing between the flowers searching for sweet nectar. Near the kitchen there is a clean, comfortable indoor area to eat, in case the weather turns for the worse.

Speaking of the weather, it's nice, sunny and warm. It's already 24°C at 7:00 am and the views are good. I can see Swayambhunath (The Monkey Temple) off in the distance, perched upon its massive hill overlooking the Kathmandu Valley. Lakpa comes to join me and we have a nice breakfast as we talk. Some fried eggs, toast with strawberry jam and peanut butter, fried potatoes, tomatoes, and mango juice.

Just after we finish eating, my personal guide for the expedition, Pema Sherpa, joins us. After a few minutes of introductions, we leave the hotel to get the rest of the gear that I need on the mountain.

After leaving, we walk past a new hotel still under construction, enveloped in bamboo scaffolding. This catches my interest—I'm a journeyman scaffolder at home, and I've never seen bamboo scaffold before. It appears to be poorly built, bending at the bottom and snaking its way up to the top of the building, but bamboo scaffolding is supposed to be the strongest. I take a couple pictures to show to the guys I work with back home.

The streets here are narrow, and many are simply compacted clay or mud. Little stores and shops line patches of pavement, with many of their wares displayed outside. All the stores are about two feet above road level, with steps the full length of the storefront leading up and into the store. This is to keep the stores from flooding during the heavy rains of the monsoon season from June to August.

It only takes us ten minutes to get to the Chhiringma Trekking Shop, run by a local sherpa and his family. It's small, about eight by ten feet inside, and their walls are filled with climbing and trekking gear. They have everything you could need here, and if they don't, they can get it by the next day at the latest.

I love it here, as their service is warm and personal. When you enter, the first order of business is milk tea, served piping hot. We sit and chat a while, then choose and size my gear. I get high-altitude climbing boots, snow goggles that are compatible with an oxygen mask, down suit, and two sleeping bags (one for basecamp and a -50°C bag for the higher camps).

After about an hour, Pema, Lakpa, and I leave, taking my bag of gear through the crowded narrow streets of Thamel back to the hotel. I spent $1,200 USD on gear, but I'll get most of that value back when I return and sell it back to the store—almost like I have rented the gear. The trekking store business here seems lucrative.

Once I'm back at the hotel, I lie down to rest. About an hour later, Lakpa, Pema, and I go just down the street to the Fusion Café for lunch. As we sit on the roof under the umbrella, we watch workers across the road building a new addition to the Holy Himal Hotel that will be the new restaurant there. The young men are wearing flip flops, even designer jeans. The guy welding uses sunglasses to shield his eyes. What a difference compared to how we work in Canada! I have a nice lunch: Nepali set, which is rice with various meats, sauces, and curried green leafy vegetables, like cabbage. The Nepali set is *dherai mitho* ("very delicious" in Nepali).

Sacred Valley Home Hotel. The veranda.

After lunch I go back to the hotel to explore this beautiful building, with its marble floors and intricate, hand-carved wooden doors depicting Hindu deities. I can't even begin to imagine the man-hours put into these carvings! The hotel has an eight-foot-tall, solid white concrete wall surrounding it, and decorated metal gates for the parking lot. The front of the hotel has a semi-circular veranda with lush green plants and flowers and blue-capped, carved white pillars. I also noticed that the hotel is absolutely spotless: the young ladies who clean here do an immaculate job, as I couldn't see any dust anywhere. The office of Lakpa and Adventure 14 Peaks is also on the main floor, just beside the sitting area of the hotel's lobby.

As I mentioned before, the walls are solid concrete. I imagine this helps keep the hotel cool during the hot Nepali summer, when daytime temperatures hover around 30-35°C with high humidity. Time to retreat back to the nice A/C of my room and rest, as I'm still recovering from the travel and time change.

I lie down for about an hour, then do a little sightseeing in the Thamel district around the hotel. Just down the street, laying on a step near a store, is a golden retriever. She's a beautiful dog with a healthy coat, and definitely well fed as she is a little chunky. I stop to say hi to her and pet her for about five minutes.

I carry on, checking out the sights along with the organized chaos called traffic. It's pretty crazy around here. Traffic and people everywhere, and intersections without traffic lights. The busier ones have an elevated round white booth in the middle, with a policeman in blue uniform and white policeman's hat directing traffic with a whistle in his mouth. And then all I can say is "Holy cow!" No pun intended, as I actually see a cow casually strolling down the road.

Many people wear masks over their mouths and noses because it's so dusty here. A lot of the dust must be from the earthquakes. The roads are mostly ruined and now just compacted red clay, and the traffic stirs the clay dust into the air. There are also many buses and trucks that spew out choking black smoke.

Garden of dreams in Kathmandu.

Next, I visit a beautiful oasis like park that is encircled by a high decorative wall called the Garden of Dreams. It has a building that looks like an Indian temple, plus a nice café. There are statues, fish ponds, gazebos, many birds and fountains, as well as a bar. Facing the main building is a terraced grassy area that many people use to relax, have picnics, or meditate. There are many beautiful flowers and trees. It's peaceful and calming in here, away from the busy streets and all the honking horns outside these walls.

There are no posted speed limits here, but you'd be hard-pressed to see an accident or even any vehicles with dents in them (and I have been looking!). I think the horn-honking is partly why there are so few collisions. People here honk to let you know where they are or they are about to pass you, unlike in Edmonton where people honk mostly out of frustration or anger.

Besides cars, I see three-wheeled bicycles with decks on the back to carry goods, some stacked ten to twelve feet high. Ornately decorated bicycle rickshaws take the tourists on tours. As I walk, many locals are curious about where I'm from. People are very friendly here, and many want to talk with me so they can practise speaking English.

Also, there are more motorcycles and scooters than I can count. I notice most of the motorcycles are 125cc Honda Heroes. There are also Royal Enfield's from India, which are nice and almost sound like a Harley Davidson with a throaty *thump, thump, thump* exhaust note from the single cylinder engine.

Many of the women are beautiful, some with stunning copper-coloured eyes that are simply hypnotic. The women here dress colourfully, many in saris with intricate gold jewellery, and have skin that many women at home I think would love to have. I think this is due to the humidity and warm weather. Everywhere I walk is the fragrance of incense, patchouli and sweet jasmine. A favourite pastime here must be following the Kama Sutra, as many children are everywhere you look! There are also many dogs that seem stray, but are happy and healthy.

Everybody appears happy and stress-free. Physically, their looks range from East Asian to Indian, and all the mixes in between. On the way back to the hotel, I get some chocolate from the corner store for the ladies who clean the rooms and the hotel. They clean everything by hand, even the floors, and want to thank them for their hard work. I notice details like this and appreciate it.

Back at the hotel, I go up to the café to have a bite to eat. On my way up the stairs, I see the ladies who do the cleaning, so I go to my room to get the chocolates. I give the chocolates to the ladies and say *dhan'yavad*("thank you" Nepali) for their hard work. They thank me with their sparkling eyes and beautiful ivory smiles.

Still feeling pretty full from lunch, I don't want a big meal for dinner. I have a tasty tuna sandwich with a Coke, with great company as young Samir Giri, the hotel manager's nephew, sat with me. I enjoyed talking with him, as he is well-mannered, personable and intelligent. After lunch, I go back to my room to write in my journal before lying down again for a catnap. Hopefully tonight I will have a decent sleep, as I could really use it. I have a busy day ahead tomorrow.

Tomorrow, I will try to find an authentic Gurkha knife called a Kukuri for Richard, a friend and the general foreman at the worksite where I sprained my knee. He asked if I could locate one for him while I'm here in Nepal.

I also want to do some more sightseeing, as Kathmandu has more UNESCO World Heritage Sites and ancient temples than any other city in the world. Buddha was also born in Nepal, in a place called Lumbini. I'll also look for souvenirs for friends back at home in the wonderful little shops here—a person could spend a lifetime checking them all out. There are so many really cool things here.

Friday, September 1, 2017

How I love waking up to the morning songbirds and the roosters crowing throughout the city! I had a good sleep last night—a solid seven hours—and I feel great this morning. From my balcony, it is a beautiful and warm sunny morning. It's just after 6:00 am and already 26°C.

After a shower I go upstairs for breakfast. Lakpa is already here, so I join him. He will be busy today, so his younger brother and climbing/trekking guide-in-training, Pemba (not to be confused with Pema, my personal guide), will take me to see a couple of the temples. Many sherpa people are named after the day of the week they were born which is why so many have the same first name! Here are the names. Sunday is Nima, Monday – Dawa, Tuesday – Mingma, Wednesday – Lakpa, Thursday – Phurba, Friday – Pasang and Saturday is Pemba.

After breakfast, I go one block over to Jyatha Street to see if I can find the knife for Richard. I find a shop that sells knives, but they are the shiny, engraved ones that the tourists prefer. The authentic knives are hand-made from the leaf springs of a truck, and they need to be kept in a light coating of oil to not rust. They are also very sharp—sharp enough to shave with, even.

The friendly shopkeeper says his brother can make an authentic knife for me, but it will take one day. I ask him how much it would cost, and he tells me 3,000 Nepali rupees, or $30 USD. We seal the deal with a handshake. As I walk back to the hotel, I'm also beginning to feel that I'm finally getting used to the heat and humidity.

CHAPTER 6:
MONKEY BUSINESS

Friday, September 1, 2017

Back at the hotel, I meet Pemba Sherpa in the lobby. We hop into the taxi usually parked outside the hotel and make our way through the crazy streets of Kathmandu, listening to upbeat Nepali music all the way to the Boudhanath Stupa.

The roads are so busy, cars and motorcycles travel within inches of each other, even in opposite directions. I swear we are about to have a head-on collision a few times, but the skill of the drivers here is astounding. At first, I have my arm out the window resting on the door, but soon decide it's safer to keep my arm in the car. It's almost like drivers here have feelers on the outside of their cars, and literally feel their way through traffic. I can only call it orchestrated chaos. It works well, as you never see any accidents, much less anyone stressed out or angry. It's so busy but so casual at the same time.

There are also so many people here walking everywhere on the roads, along with lots of dogs, and I see a few more holy cows out for a stroll. On scooters and motorcycles, only the driver wears a helmet. I even see a complete family of husband, wife, and two children on a motorcycle, and only the husband who is driving has a helmet on!

The unbelievable power poles of Kathmandu.

As in other Asian countries, Kathmandu has mini three-wheeled taxis called *tuk-tuks*. The huge and colourfully painted transport trucks are like those you see in India, with loud and comical sounding horns the drivers aren't shy about using. Micro buses are so crammed full that people are hanging out the windows. I'll bet they are good at ducking in to avoid power poles.

Speaking of power poles, I truly feel sympathetic towards electricians here. Almost all the power poles down the streets have a couple of hundred wires wrapped loosely around and on them, looking as though a family of octopuses are clinging on for dear life. I'm willing to bet there is no such thing as an unemployed electrician here.

Away from the touristy areas, I also see many older buildings that suffered damage in the massive 2015 earthquake. There are piles of rubble and occasional stacks of fresh bricks that local people can use to rebuild. I understood the Nepali government received lots of money from the rest of the world and the Red Cross after the earthquake. This money was intended to help the Nepali people, but the government has still not given any of this funding out to the people who need it.

As poor as many of the people are here, I truly believe they are the richest in spirit, soul, and compassion. No matter what their skin colour, religion, or beliefs, they support and help each other survive. The rest of the world could learn so much from the people here.

The banyan trees here are old and massive, as well as beautiful with their lush green foliage and twisted trunks. Some of these trees which are considered gods in Hindu and Buddhist culture. These trees are so big that next to them, our large elms and maples in Canada would look like shrubs.

We arrive at the UNESCO World Heritage Site Boudhanath Stupa, whose large mandala is one of the largest spherical stupas in the world, standing thirty-eight metres tall. Built around the fourteenth century, this stupa supposedly entombs the remains of Kassapa Buddha.

Boudhanath Stupa is one of the most visited sites in Kathmandu. Along with Pemba and me, there are many others circling the stupa clockwise and spinning prayer wheels by hand. In Hinduism and Buddhism, it's proper and respectful to walk around temples and stupas clockwise at least three times.

Boudhanath Stupa entrance.

When originally built, the stupa had its own little village encircling it. This is where many Tibetan peoples originally settled in the valley. Now Kathmandu has grown and spread out to surround the stupa and its neighbourhood. Directly around the stupa are shops, rooftop cafés , and vendors selling their wares. There are also hundreds, if not thousands, of pigeons. You'd think there would be many droppings on the ground or the stupa itself, but I don't see any. It's almost like the pigeons are reincarnated souls and respect these sacred grounds.

After driving to the other side of the city, we arrive at Swayambhunath (The Monkey Temple). This UNESCO World Heritage Site is perched high atop a hill that overlooks the Kathmandu city and Valley. At the entrance, two eight-foot-tall Buddha statues sit on either side of the walkway. Just past these statues are 365 stairs that lead to the top, where the temple and stupa sit. On our way up the steps, we pass some Nepali women brightly dressed in traditional wear, carrying baskets full of bricks on their backs with head straps to support the weight.

Once we get near the top, I can see some men high up on a scaffold using these bricks to rebuild one of the temples, which suffered damage in the big earthquake. I'm now soaked in sweat from the temperature, humidity, and climb up the steps. At the top, I see a beige dog walking past the stupa with some strange growth or lump on its back. Looking closer, I realize it is a small monkey riding the dog. *Damn,* I think as I fumble to get my phone out to take a picture. But it's too late: the pair have disappeared around the stupa. *Talk about monkey business!* I think to myself with a laugh.

This temple is a sacred Buddhist home of spiritual monkeys. Brass statues of ancient deities as well as Buddha are all around the stupa, and the monkeys (macaques) are absolutely everywhere. They are in the trees, on the ground, on the stupa, swinging on the ends of prayer flags. Some are feeding their babies—I guess breastfeeding is allowed in public here!

There are also about twelve dogs up here, happy and well fed as usual for Kathmandu. A dog's life would be great here! I already love the people here, just for how well they care for animals. We spin prayer wheels as we walk around the stupa, and there are also more vendors here selling their handcrafted wares, much of it intricate jewellery.

When Pemba and I come around to the other side, we see a Buddha statue standing on a platform in the middle of a pond circled by a brass railing.

A brass bottomless catch pot is at Buddha's feet, over the water. People are tossing coins, trying to get them in the pot for blessings and good luck. About fifteen people are cheering each other on as they toss their coins. Pemba and I trade a couple paper bills for coins and try our luck. It was fun with everyone cheering, and we both got a few in the pot.

Buddha statues of the Swayambhunath entrance.

Breastfeeding allowed in public at Swayambhunath.

Brass guardian statue of Swayambhunath.

Buddha at Swayambhunath.

Buddha blessing pond.

By now, I must sound like a cliché tourist trying to pump up a destination. But unless you come here and see all this for yourself, it's hard for me to find the words to describe how wonderful it is here—especially considering most of this was built over a thousand years ago.

Before heading back down the stairs, we go over to a viewing deck overlooking Kathmandu. What a view! The lush green valley, the multicoloured flat roof buildings, and the jungle-covered foothills around all sides. With the mighty Himalayas as a backdrop, it looked just like a painting.

Back at the hotel's rooftop café, I'm enjoying a Coke and talking with Lakpa Sherpa about the expedition. I love the Coca Cola here because it still comes in glass bottles. Not only that, but it just seems to taste better: it's colder and inexpensive at seventy cents per sixteen-ounce bottle. I've noticed that everything pre-packaged here, like chips, chocolate bars, cookies, and soft drinks, comes from India.

In the lobby, I meet with an older gentleman whose name is Jeevan Shrestha. He is here to interview me and record my information for my attempt to climb Mt Manaslu for the Himalayan Database with Reuters

News Agency. American journalist from Chicago named Miss Elizabeth Hawley started the database. Miss Hawley moved to Kathmandu in 1959 and never left.

Rumour has it that Miss Hawley and Sir Edmund Hillary had a bit of a romance between them. Sir Edmund was the well-mannered, modest bee-keeper from New Zealand who famously climbed Mt Everest with sherpa Tenzing Norgay—the world's first people to reach the summit. Sadly, we lost Miss Hawley to pneumonia in January 2018 here in Kathmandu at the CIWEC Hospital Pvt Ltd. She was ninety-four.

After the interview, Lakpa and I go to the Fusion Café for an early dinner. I had chicken momos, which are chicken-filled dumplings that come with a to-die-for curry dipping sauce. They are so yummy! Momos also come with other fillings, too. I believe I have found my new favourite food.

Saturday, September 2, 2017

I slept well last night; the bed in my room is so comfortable. I can't remember what they were, but I had some strange dreams last night. The good thing is I had some deep REM sleep.

The mornings in Kathmandu never disappoint, with their warm sunshine and birds. I eat breakfast with Lakpa, and Pema Sherpa also joins us for the meal and to assist me with packing my gear. Today is the day I will pack all my things for the expedition.

I still need to purchase some fleece wear as well as an umbrella. I will also need to adjust the high-altitude boots to my feet with thick wool socks on. They have an inner boot and an outer semi-rigid boot with adjustable lacing that you pull and secure with Velcro.

Later this afternoon, all of the team members will be meeting for a meet and greet and Q&A session with all of our sherpas at a restaurant not far from here. This is also when we will receive our climbing permits from the Nepali government. Not everyone will be there, as some team members are flying to Samagaun rather than trekking there.

I'm back in my room after shopping with Lakpa and Pema. I got two sets of fleece wear, flip flops, an umbrella, and a new thick pair of wool socks that I will only wear for climbing to the summit for maximum loft and insulation. Now I understand why Pema needs to help me with packing, as one bag will be just trekking gear and the other bag will contain my high-altitude and summit gear.

Pema is an experienced sherpa and I enjoy his company. Like me, he has a permanent furrow between his eyebrows, so we both look kind of angry even when we are not. The packing all done, Pema leaves and I have time for a rest before we go to the restaurant for the meeting.

I just got back to my room from the meeting and it went well. On the team is Jitesh Modi (India), Soham Saigonkar (India), Yasushi Kawahara (Japan), Nadev Yehuda (Israel), Laurent Perruchon (Italy), Riccardo Bergamini (Italy), Mathew Eakin (Australia), Madalina Condrea (Romania), Bogdan Velev (France), Alessandro Corazza (Italy), Sergio Zigliotto (Italy), Catalina Castro (Spain), and me from Canada. Time for some sleep now as tomorrow will be a long day on the road.

CHAPTER 7:
HIGHWAY OF TRAGEDY

Sunday, September 3, 2017

It's my last day in Kathmandu. I slept well and woke up at 6:00 am. I grab a quick bite for breakfast then load my bags into the van to go pick up the other members.

We are on our way out of Kathmandu now. With the driver, there are eight of us in in the van, along with all our gear: Mingma Tenzi Sherpa, Pema Sherpa, Jitesh (Jite) Modi, Soham Saigonkar, Sergio Zigliotto, Laurent, (Lollo), Perruchon and me. The rest of the team will fly to Samagaun by helicopter and meet us there.

As usual, the weather in Kathmandu is beautiful this morning. Just before leaving the city, we drive through an area that was hit especially hard by the 2015 quake. I can see piles of rubble that were once buildings. The roads here are also dusty and in much need of repair. The soil here has a lot of red clay in it, which explains much of the dust and why many buildings are made of brick. Once we are out of Kathmandu, the dusty roads become nice smooth pavement like the highways at home.

The countryside here is lush, green with fields of corn and terraced hills for growing rice and vegetables. The farmers' houses are nice-looking and colourfully painted. They are large, almost like mini apartments of four to five storeys.

After about an hour and a half on the road, we stop at a small teahouse attached to a farmhouse with chickens and rice paddies outside. We have *dal bhat* (rice and lentil soup) for lunch, along with fresh chicken, rice and curry. It's so tasty that I could stay here all day, just eating.

The highway is getting busy now as we begin to climb into the hills. There are many tour buses and those brightly painted trucks with their loud, unique horns. Those horns would scare the crap out of anyone within twenty feet who wasn't expecting it! Many of the trucks are ornately decorated, some with funny sayings on them. I even see one with Bob Marley painted on it.

As we climb into the hills, the highway becomes narrow, with barely enough room for vehicles to pass without clicking mirrors. Tin shacks are set up sparsely along the road, selling Tuborg beer, snacks, and Coca Cola, of course.

Suddenly it all comes to a grinding halt. You wouldn't expect gridlock on a highway! Slowly we manage to move up the road a bit, then I can hear it: a god-awful crying, screeching sound, with a whimpering between the cries that makes my soul cringe with agony. It is just heartbreaking to hear.

The van moves up the road, and oh, my God! In the middle of the road ahead is this poor dog. Still fairly young-looking, he's been run over—not just hit, as his rear end is crushed. As we get closer, I can see part of his pelvis sticking out through his fur. The poor thing is crying out in pain as he tries to get off the road. Only one of his front legs seems to be working, and all he can do is go in slow, excruciating circles as he cries out in agony.

All of our hearts sink, mine so much so that I ask Mingma Tenzi if I can get out, find a big rock, and put the dog out of his misery. It's obvious he's done for, and I see no need to prolong his suffering. Mingma says that I could get into trouble if I did this, and it would be best to let nature take its course. As I do not yet understand the laws or culture here, I agree and stay in my seat.

As we're about to drive past the dog, one of the truck drivers scoops him up. The cries of pain are unbearable. I keep watching as we drive on, and see the trucker placing the dog in the tall grass on the side of the road to die alone. As the driver walks back to his truck, he wipes tears from his face.

We're now at the village of Besisahar for the night, in the Annapurna Conservation Area and not far from the famous Mt Annapurna. We stop at the Throungla Guesthouse Hotel and it's quite nice.

Our hotel in Besisahar.

I go to use the washroom and I see the toilet: it's a squatty potty, which I have never used before. For those of you who don't know, it's a toilet bowl built into the floor. You squat over it to do your business, and you flush by pouring water from a bucket into the toilet bowl. *These are gonna take some getting used to,* I think. *At least there's toilet paper!*

This quaint little hotel is owned and run by a family; the innkeeper and his wife have six children, two boys and four girls from toddler to fourteen. The family is so warm and welcoming. I brought many Reese's chocolate peanut butter bars and I give a couple to the children. They love them and now I am their best friend. For dinner, we have tasty homemade pizza. Even Sergio and Lollo, the Italians, gave it two thumbs up. Now that's saying something!

Monday, September 4, 2017

Jite, Soham, and I share a large room with four beds. Since it's so warm here throughout the year, our rooms have windows without glass. Instead, there are shutters with slats and decorative flat steel bars. As we are trying to sleep, little buzzing vampires try to attack us. We finally get rid of them and plug in a mosquito repeller. Now all I can hear are the crickets, which are wonderful to fall asleep to.

I will never become tired of waking up to the crowing of roosters. I had a good sleep and today our porters will join us for the rest of the trip.

When we arrived yesterday, the porters met us here and pulled our gear bags out of the van. Today we will be transferring to tough four door jeeps with short truck beds called Mahindras. These vehicles are tough enough, in fact, that both the Indian and Nepali army use them. There is a good reason for the jeeps: many of the roads are stream or river beds, some dry and some flowing. We will also be driving on rock-strewn trails and fording some rivers. Some of the roads are narrow, blasted out of the sides of sheer rockfaces with a drop of at least a thousand feet to the rocks and river below, just like you see on Discovery Channel. Many people die on these roads every year.

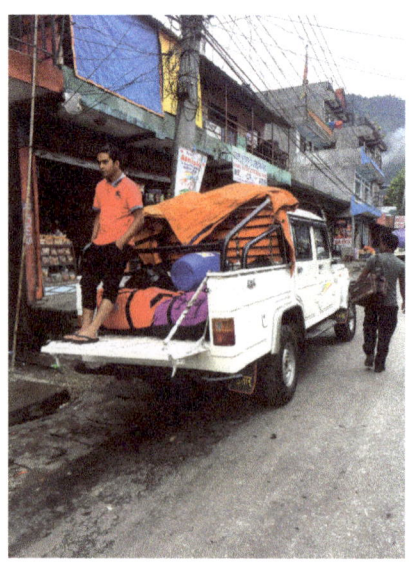

Loading up the Mahindras.

We left Besisahar the sherpas and porters riding in the box of the jeeps with our bags. Inside, we are comfortable and listening to Nepali music, which I like very much. Not far out of town, we are already driving up a flowing streambed that also serves as a road, and it's a slow, rough ride over the rocks.

The road smooths out a bit and we stop for a pee break. Lollo, wearing shorts, steps into the bushes to relieve himself, and when he comes back out, he has leeches hanging off his legs and dining on him. He gets them off quickly and we have a good laugh.

Back on the road, we drive through a tunnel at least a kilometre long, travelling through the base of a foothill. About ten minutes after the tunnel, we come to a grinding halt again. Just ahead of us, there had been a landslide right across the road, and there was no way we could have driven over all the debris, rocks, and dirt. Lucky for us, on the other side of the slide are two jeeps with some locals and

Landslide across our road.

monks onboard. We end up simply trading jeeps. It only takes about an hour to pack all our gear and supplies over the debris, then once again we are on our way.

We start to gain considerable altitude as we drive up what I would call goat trails littered with fallen rocks. It's a slow, rough ride in first gear. I'm sitting on the left uphill side, next to the back door. On the right side is a sheer drop of at least five hundred to a thousand feet. Jite is sitting next to this door, and a couple times his eyes become as big as saucers because the jeep's wheels were within inches of dropping off the side.

It might be a good idea to keep my hand on the door handle, as I am completely ready to bail out if need be. Sitting beside me, Soham sees my hand on the handle. He leans towards me, ready to jump, too. Jitesh tells us he can see a couple jeeps and busses wrecked at the bottom near the river.

Narrow gorges on the way to Upper Dharapani.

Rocky Himalayan roads.

The scenery here is nothing short of spectacular. Deep Himalayan gorges with lush, bright green jungle clinging to their steep sides, as cascading waterfalls plummet hundreds of feet into the valleys and raging rivers below.

We stop for some food at the Waterfall Restaurant, directly under an over-hung rockface. Across the valley, mist from a beautiful waterfall glistens in the sunlight and forms rainbows as the water gracefully plummets hundreds of feet into the torrent below.

The view from the Waterfall Restaurant.

We drive through a couple rivers and streams, and then over the rapids on a rickety log bridge. Some of the logs pop up and hit the bottom of our jeep as the tires roll over them. I think for sure we are going to end up rolling over into the water! We then climb up another steep trail and onto one of those narrow roads blasted out of the side of a sheer rockface. It's like a three-sided tunnel, as it has a bottom, a left wall, a roof, but no right wall. It just drops off into thin air of the abyss.

After we drive around a corner, there's so much water rushing down at us as it plunges off the side to the valley below. Ahead I can see that the road is the landing point for a waterfall cascading down the rock from at least a hundred feet above. I roll up my window as we drive through the waterfall, and all I can hear are the sherpas and porters shrieking and hollering in the back as they receive a cold shower. It sounds like they're having too much fun back there as I laugh.

We arrive at the village called Dharapani, which is bisected by another deep gorge with a raging white-water river and waterfalls rushing over boulders the size of houses. The stone houses on the other side are accessed by a 300-foot suspension bridge that hangs precariously over the gorge carved out by the water. As we drive cautiously up the narrow street to our hotel, I see pack mules along the side of the road that are not loaded, some with feed bags of corn hanging on their heads. They must be resting for the night.

Me on the expanse of the suspension bridge at Dharapani.

Now we're home for the night at the Kangaroo Guest House and Restaurant. I like it here. It has two levels; our rooms are upstairs and accessed by a spiral staircase, and yay, it has a squatty potty. I have my own good-sized room, with glass-paned windows facing the river, which is also easy to hear. The porters unload all of our gear from the jeeps because tomorrow we will be on the trail, trekking for about seven to nine days.

Okay, I am all settled into my room, Jite, Soham, and I are going for a walk and do some exploring before dinner. We leave the hotel, then walk across the suspension bridge. It's a long way to the other side, and roughly forty feet above the torrent and boulders below. As I am following Jite and Soham, I'm taking some video to show the length of this narrow bridge. Once we get to the other side, we walk between beautiful stone houses with stone fences. Corn grows behind the fences, along with many flowers, such as rhododendrons and other colourful varieties that fill the air with their sweet fragrant scents. This reminds me of springtime visits to the greenhouses with my mom and dad to buy flowers for our yard.

We then come across a couple repairing the stone steps that lead up to their house. They say hello to us and we reply with, "Namaste." They ask where we are from; we tell them and add that we are on the way to climb

Manaslu. They wish us luck as we continue on. This would actually be a nice place to live, I think.

Walking out from between the houses, we reach an open trail. Ahead I see three schoolboys with their small backpacks and books in hand. *Isn't that strange?* I think as I see one of the boys bending over and rubbing plants between his hands. Then I see what the plants are: it's marijuana, and he is making himself hashish. Jite talks to him and he shows us a big wad of hash in his hand.

The weed is growing everywhere. It appears to be the Sativa strain, whose plants are short and bushy with sticky buds and leaves. The resin is what the boy is collecting. The Sativa strain makes you want to be active, whereas the tall Indica strain is better to smoke to relax or use before sleep (but I wouldn't know anything about this). We leave the boys behind, and no, we didn't smoke anything!

Up ahead of us is a white stupa, about twelve feet tall and six feet across at the base. I take a few pictures with my phone. We cross the river again over a series of waterfalls, on a bridge big enough for jeeps to drive over. Beside the large bridge are the remains of the original suspension bridge. Only a brick base with built in stairs are left, covered in deep green moss.

Once on the other side, we turn back towards our guesthouse. It's all downhill travelling back. As we follow the trail along a steep embankment beside the river, we come across six goats grazing in the tall grass on the side of the hill. One of them is eating the marijuana growing there, and looks like he's having a great time jumping all around like he is a kid again.

As he disappears in and out of the long grass, I say to Jite and Soham, "Look, a camouflage goat!"

About halfway back to the guesthouse, a porter is walking uphill towards us with an immense load on his back, held up by a head strap. His load looks like it weighs at least forty or fifty kilos. It amazes me how truly strong the people of these higher altitudes are.

Back at the guesthouse, we relax in the dining area and meet the inn-keeper's five-year-old daughter, Sarah. She is a bright and personable little girl with good manners. Her little brother, a toddler, is wandering around, but I don't get his name. Sarah is sitting with Jite at a table near the entrance to the kitchen where her mom is preparing dinner for us. Jite is reading one of her books with her and it looks as though they are having fun. I'm sitting at a table on the other side of the room by myself as I write in my journal the days activities.

Porter passing us on the trail.

Little Sarah spots me and takes an interest in what I'm doing. She then comes over, plops herself right down beside me, and wants to draw a picture in my journal. I can't refuse, so I give her my book and pen. She then draws a nice picture of herself and her brother.

Her brother joins us and stands beside the table, then looks at me as if to say that he wants up. I pick him up, sit him on my lap, and let him scribble in my journal to his delight and I shoot some video of this.

The children leave and we have dinner. I have tasty and filling spaghetti and cheese. After dinner I sit with the boys for a while and then go to my room. Time for some sleep now, as I am tucked into bed with a full belly, listening to the river outside.

CHAPTER 8:
HIMALAYAN VULTURES AND DEATH WARMED OVER

Tuesday, September 5, 2017

Awake now! It's about 5:30 am and my sleep was very good; I actually feel well rested. Gonna get my things ready to go then have some breakfast. Mmmm, scrambled eggs and chapati bread with strawberry jam! (Chapati is a flatbread also known as roti)

My new little friend Sarah joins me, sitting right beside me to eat her breakfast. Our porters are outside, leaving ahead of us with our gear bags on their backs. Some are sherpas in training, working their way up the ladder, or mountain, so to speak. They are small of stature and thin, I think no older than twenty, but strong, happy, and helpful. Most wear only flip flops, and a couple have running shoes. Some of them are even wearing designer jeans.

About a half-hour later, we say our goodbyes as we head out with our packs on and trekking poles in hand. I give little Sarah a Reese's bar before I leave, and she stands there with her mom waving goodbye to us.

It's a very nice morning, with only a few scattered clouds. We begin to follow the deep turquoise Marsyangdi River, which runs fast and has many small waterfalls. It reminds me of the Maligne River in Jasper National Park back home with is greenish turbulent waters. On the trail ahead is a mule train, packing supplies for the village. I can hear the mules' bells ringing as we approach each other. By the looks of it, they are mainly packing bags of rice and corn. We come upon their drop point, where there are many bags and drums staged beside the trail.

Now we have caught up with our porters, and we slow down a bit to keep pace with them. One of them has a Bluetooth speaker proudly playing Nepali music. It's uplifting and brings smiles to everyone's faces. I recognize some of the songs from Priyanka Karki's music videos, and quite enjoy them.

We begin to climb some steep switchbacks that bring us out of the valley quickly. About halfway up, one of the porters sees a goat trail that heads straight up the slope through the trees and brush. But with the weight he's carrying and his flip flops, it's too difficult even for him, and I think Mingma Temba Sherpa talked him out of it as well. The porter then stays on the switchbacks with us.

It's getting slippery on the trail with some morning dew left in the shady areas. We need to watch our footing and avoid little mounds of what looks like dollops of creamy chocolate pudding. Mmmm, chocolate pudding! Oh wait—it's mule poop.

As we gain height and move into the full sun, the temperature and humidity are rising. Along the top of the valley I continue with the river far below on my left, the trail straightens and levels out. Ahead of me I can see four young girls wearing their school uniforms. Which consist of blue knee high socks, blue skirts and white shirts. I stop to pull four Reese's chocolate/peanut butter bars out of my bag and give them to the girls. They all give me huge smiles and say, "*Dhan'yavad* [thank you]," and continue on their way.

Mingma Tenzi Sherpa is now leading the way. I'm behind him as we enter a small farming village in the Gandaki area. It's another scenic village with stone houses that have stacks of firewood eight feet high along some of the walls. Some walls are brightly painted in sky blue and bright pink. As we walk through, we ascend flights of two to three stairs that are about twenty feet apart, with flat stone walkways in between to create an easy and gradual rise.

On the front concrete veranda of one of the houses, a few hundred fresh baby potatoes are piled in the sunshine. Not more than ten feet away lies a caramel-and-white dog basking in the sun. Fittingly, his breed appears to be Himalayan mountain dog. As we pass by, he raises his head. I say hello to him, and he just lays his head back down.

Out of the village now, we are still following the river. It's very warm considering it's just after 10:00 am, and the temperature is already a sweltering 32°C. Mingma Tenzi Sherpa stops and doubles back to talk with Pemba Sherpa. Pemba is a big man and an experienced sherpa, although you wouldn't think he is an accomplished climber if you saw him on the street. You can't judge a book by its cover, as he has much experience on Mt Everest. All of the sherpas are great to be with, always smiling, happy and helpful.

I'm now in the lead as we follow the raging river below. Behind me are Sergio and Lollo, the Italian climbing machines. There are so many beautiful waterfalls throughout this valley, cascading down hundreds of feet and creating shimmering vivid rainbows in the sunlight. It's simply heavenly.

It's really hot out, so I begin to run ahead so I can stop in the shade somewhere, rest, drink some water, and take off my shirt. I find a big tree on a small rise beside the trail and stop there. I finally cool down, and about twenty minutes later I see Sergio and Lollo approaching. Once they are close enough to hear me, I say to them in my best Italian accent, "Heya whatsa goin onna? Youa nadda eata nuffa pasta, ha, ha, ha!" We all laugh.

We walk about ten minutes and come upon four-foot-tall stone fencing along both sides of the trail. Corn is growing behind the fence, and I see some small stone farmhouses ahead. A woman is standing outside one of these houses, holding her little girl in her arms. The girl looks about three to four years old, and they both have rosy cheeks from the sun.

The woman watches me approach and says, "Namaste."

"Namaste," I reply. Then I take off my pack, pull out a Reese's bar and hand it to her for her and her girl. "Photo?" I say, and she nods yes. I take a nice picture that I think Reese's would love to use for an ad, and continue on.

It's now almost noon; we stop on the top of a hill at a guesthouse and restaurant called Top of the Hill Guesthouse. It has a remarkable view of the valley and a few waterfalls. I walk inside the dining area and am stunned by the handcrafted woodwork in here. Amazed, I begin to take pictures of all of it. The details are so intricate. Even the tables and chairs are all hand carved. This whole place is a work of art.

The guesthouse is run by a woman and her daughter, who are in the kitchen preparing food on a wood-fired stove (like all stoves in the back country of Nepal). Outside is a nice yard with well kept grass and a table with chairs. The rooms are on the second floor. Hanging over the wooden railing is the day's laundry to dry in the breeze and sunshine. Lunch is a yummy omelette with goat cheese and some perfectly spiced boiled potatoes.

We are back on the trail; the terrain is becoming more rugged as we gain altitude. We are entering a dense forest, and inside this forest is another type of dense forest: marijuana, Indica strain, and lots of it. Some of the plants are eight feet tall. No wonder the hippies of the sixties loved Nepal, as marijuana seems to grow nearly everywhere here. I did pick a nice juicy bud and pin it to my chest strap like a lapel pin.

The terrain flattens out, and the marijuana forest gives way to an old-growth forest. Tall grasses grow among these ancient trees, which are at least two hundred feet tall whose branches are sparse and six to eight feet across at their base. They remind me of the California Redwoods. At our next stop, I ask a couple locals about these trees, and they tell me they are hundreds of years old, and some may have even been alive at the same time as Buddha.

Now we trek into more jungle type of terrain, with lush green foliage and smaller, bushier trees. It has also become eerie, as fog has rolled in and this dense forest looks like a horror movie set. It amazes me how the terrain and flora here changes so drastically in such a short distance. It's nice in here, somewhat cooler with lush grasses, ferns with huge leaves, and many beautifully coloured flowers. And butterflies—some are the size of small birds and look like someone hand-painted them in all the colours of the rainbow. I see a small lizard that looks similar to a gecko, but larger. I had no idea that places like this even existed; absolutely beautiful, like Buddha's garden.

We're now leaving the jungle and are walking back down towards the river. I'm with Jite and Soham beside the river, and I stop to take the photos of some of the flowers. Meanwhile, the guys get ahead of me and out of sight. I follow, but I'm alone as I come to a fork in the trail.

I stop and wonder which way to go. To the left, it seems to climb back up into the trees, and to the right, it stays along the river. I decide to continue along the river, which was a good choice, as I soon catch up with Jite and Soham.

We walk up a short hill and through some bush, then it opens up to neatly kept grass, as well as nice mini cabins and two larger buildings. I breathe a sigh of relief. Our next stop for the night: the Seven Sister Hotel and Restaurant in the Karcha district.

Here the river we've been following joins up with another smaller river. It's just after 4:30 pm, and the day has grown cloudy and cooled off considerably.

I grab the cabin on the end for myself. Behind it is a garden with all sorts of vegetables. The cabin is blue and has stained wood with a tiny porch just large enough for two chairs, with all sorts of flowers growing around it. Inside it's cozy, six by eight feet with two beds.

My quaint little cabin at the Seven Sister Hotel and Restaurant.

Once I'm unpacked for the night, I go back outside to have a look around. I meet three of the seven sisters who own and run this hotel, friendly young ladies with big welcoming smiles.

Beyond the building that contains the ladies' quarters and the kitchen/office, I see a fenced-in staging area where the pack mules are. The mule driver is unloading the mules for the night and putting their corn feed bags on them. I have heard the mules are packing supplies for a Russian expedition that is two days behind us; they lost a day waiting for the landslide to be cleared off the road. They weren't as lucky as we were, with jeeps on the other side.

I'm video recording the mules as they are offloaded, and a straggler joins the others. He grabs a small branch off a tree and chews on it as he casually approaches. I learned from their driver that mules do this when they are hungry. The trail we were on continues past the mule corral and crosses the smaller river over a small steel bridge draped with brightly coloured prayer flags. We will cross this bridge tomorrow morning when we leave.

After a nice dinner, I sit around the wood stove chatting with some of the porters and sherpas. I learn that most of them work on Mt Everest every year, working Mt Manaslu in the fall to make extra money to support their families and pay for their children's education.

Well, time for bed now! *Subha ratri.*

Wednesday, September 6, 2017

Sitting at the base of the tree with it's insides burnt out from a lightning strike.

After a great sleep in my little cabin, I get up, have a quick breakfast and I am back on the trail at 7:30 am. We cross the bridge then trek through another forested area with some fairly large trees and bush. This looks like a bad spot to be during electrical storms, because a couple of these tall trees look like they've been hit by lightning.

As I sit at the base of one of these to rest, I ask Pema Sherpa to take a picture of me. Hit by lightning that charred its insides, the tree is still living, with lush green foliage and unburnt bark. Where I am sitting at the base, some of the bark has blown off, and you can see its interior is burnt to charcoal.

After resting, we set off. The trail is all uphill and steep; we are gaining some decent elevation now, and it's obvious we are climbing a large foothill. It takes a while, but we make it to the top, then rest again and drink. I see another drastic change of scenery over the other side of the hill. It's a moraine-filled valley where a glacier long ago carved out the side of the foothill, and it looks desolate with rock, boulders, moraine, some moss, and small shrubs. A glacial stream with its milky-green silty waters rushes to the valley below. I can see that our trail meets with a sun bleached wooden footbridge, about three feet, to cross those icy cold waters.

On the way down, I'm following big Pemba Sherpa on the steep switch-back trail. The terrain is so barren, it looks like we're on another planet. After crossing the bridge, we trek a couple kilometres to climb out of this valley, reaching a wide-open meadow with many shallow streams whose crystal-clear waters meander all about. Short grass, flowers, and moss is everywhere. All around us, the steep rockfaces of the Himalayas stretch high up into the clouds.

Ahead I see a few buildings, a couple corrals, and a few grazing mules. Pema Sherpa tells me this is Bhimtang and we will be staying here for the night. Too bad about the clouds, as I'm told the views of the surrounding mountains are excellent. In good weather, you can see Manaslu, Lamjung, and Himlung peaks.

The temperature is definitely cooler now, as we are at a much higher elevation and not far from the glaciers that reach deep into the valleys below. The air is thinner now, and it's a little harder to breathe. It feels as though we will get rain tonight, as we're getting socked in with the cloud now and the humidity is rising.

We're staying at the New Tibetan Hotel. The buildings here are all made of stone on the exterior, wood on the inside, with bright blue metal roofs. I go to my room, and like my past rooms, it's only big enough for two beds one on either wall with a walk space in between. The door is secured by a big shiny padlock and key, like what you'd imagine seeing on a pirate's chest.

I go through the usual routine and unpack for the night, then come out and have a look around. Near the kitchen are two older men with a bowl of freshly picked garlic as they prepare it for the kitchen. At home I eat garlic every day, and I asked them if I could have a few. They gave me five cloves, one of which I ate immediately. It was good and hot.

Out of the kitchen comes a stunning Nepali or Tibetan woman with raven hair braided almost to her knees. She appears to be in her mid thirties, and I find her to be friendly and personable. Maybe I should consider marrying a nice Nepali woman.

We gather in the dining room; I'm quite thirsty, so I decide to try the sea buckthorn juice. Mmmm, it's really good! I have another couple of these in a large mug with my food. For dinner, I try yak meat for the first time, but I find it a little too chewy for my liking. Luckily, dinner also came with macaroni and cheese. Shortly after dinner, I retire to my room to sleep as I'm quite tired.

Thursday, September 7, 2017

I had a decent sleep but am awakened by loud, heavy rain on the metal roof. I fall back to sleep after fifteen minutes and wake up at 5:30 am. I look out the window to see that the sky is clear and dawn is approaching. Thank God—I don't like getting wet and cold in the rain. I only like rain overnight, and preferably not as loud. I get ready, eat, and we head out on the trail by 6:00 am.

This will be our longest, most difficult trekking day: we will be travelling twenty-two kilometres up and over the Larke Pass at 5,106 metres elevation. This is higher than I have ever been before, other than flying. I've heard this section is a suffer-fest, as the elevation gain is quick.

After about a half-hour on the trail, we are now gaining elevation steadily. The sky is clear, and the vistas are spectacular—I can see all the snow-capped peaks that surround us.

This is incredible, I am loving this! If only my mom and dad could see this too. I'm hoping that they are with me and can experience this through my eyes.

We come to a teahouse at a spot called Phedi. It isn't much: a couple of stone buildings with tarps for a roof that is barely above your head. Prayer flags printed with sacred scripture blow in the katabatic breeze that comes down off the glaciers, releasing their prayers to the sky as they flutter about. I try to go inside this temporary-looking teahouse, but the smoke from the wood-fired stove is too much for my lungs and eyes. Instead I rest outside in the warm Himalayan sunshine.

Teahouse at Phedi and an amazing view of the Himalayas.

About a half-hour later, we get back on our feet and are on our way up to Larke Pass. The rocky slope that goes up and over the pass looks to be a long way up, with plenty of switchbacks to ease the trail's steepness. I start to feel some of the effects of the altitude gain, but it's not too bad. Stopping every once in a while to catch my breath is working quite well. I'm also breathing in deeply with an open mouth, then out through pursed lips to build pressure in my lungs and force more oxygen into my bloodstream.

About halfway up, a driver with a mule carrying a woman passes me. I recognize her from Bhimtang. Her husband appears not long after her, still on his feet. He explains that his wife became ill from the altitude, and he hired the mule and driver to bring her up and over the pass. I guess it cost him a fair bit. Lucky girl!

6

What I don't like is that Pema Sherpa, my personal guide, will go up ahead of me about two hundred feet. Then he stops, stands there, and just watches me as if to say, "Hurry up!" At least that's how it feels to me. He does this all the way to the top, and I feel he is rushing me. I am paying him, and if anything, he should be staying with me at my pace.

Anyhow, on and up I go. Two thirds of the way up, I see some bones on the side of the trail. Once I'm close enough, I can see they're the remains of a mule. The bones are somewhat scattered so it looks as though something snacked on the mule after it died.

Then something in the sky catches my attention. "Holy crap," I say. A bird that seems to be the size of a Cessna 152 is soaring above, then it goes out of sight over the pass. Maybe I'm lacking oxygen and hallucinating, as I can't believe my eye's and what I just saw!

As I climb higher, I see more of these monstrous birds soaring on the upslope winds. One of them flies about forty feet above me, and I can now see that it's a Himalayan vulture, also known as a griffon. How cool is this! These vultures normally have wingspans of eight to ten feet and have been known to fly away with full-sized goats. Now I have a pretty good idea of what happened to the mule after it died. I am in awe of these massive, graceful birds soaring so effortlessly on the breeze.

I finally make it to the top of the pass and the view up here is cool. I feel I can reach out and touch the snow-capped peaks. The summit marker reads, "Larke Pass, 5,106 m. Thank you for Manaslu visit, see you again." I take a couple pictures and then begin my trek down the gradual descent on the other side.

I get about two hundred metres down and start to feel a little woozy—then it hits me like a ton of bricks! I suddenly feel very ill: I'm nauseous, my pulse is racing, and I feel like I'm about to crap my pants. I need to sit down before I black out, I feel like death warmed over! I'm so thirsty, my tongue sticking to the roof of my mouth. I shouldn't be dehydrated, as I've been drinking lots of water with electrolytes, and the last time I peed my urine was light in colour indicating good hydration.

I sit there, trying to recover for roughly ten minutes. Pema finally comes back to me and asks what is wrong. I tell him, and he gives me his water as I have already finished mine. He also gives me a Snickers bar.

I ask him what happens if I pass out and if he has any way to summon help. He tells me that Pemba Sherpa is behind us and has a satellite phone with him. This Pemba Sherpa is not the same as Pemba Thendu Sherpa who was taking me to the temples in Kathmandu. This Pemba is a bigger, older man with many years experience. After another five minutes or so, Pemba comes up to us. He asks about my condition and symptoms. I tell him, and that I'm slowly improving now.

Pemba says that he doesn't think it's altitude illness, since I haven't experienced any headache. Instead, Pemba says it sounds like something affecting my stomach. He asks me if I drank any sea buckthorn juice when we were at Bhimtang.

I think it's weird that he asks this, and I proclaim, "Yes! I had a few mugs of it with dinner last night."

With a smile as he nods his head up and down, Pemba says, "That's your problem!" He tells me that they don't use boiled or treated water when making the juice, and this sort of illness has happened to someone else before.

He then asks if I'm able to walk for a bit if Pema carries my pack for me, explaining that we should get to a lower altitude and there is a nice resting place about twenty to thirty minutes ahead. I agree and give my pack to Pema. I slowly make my way down to the resting spot, where there is some grass and moss to sit on.

Once I'm resting, Pemba gives me an apple to eat, along with a can of mango juice and more chocolate. He doesn't have much water, and Pema or I don't have any left either. Luckily, I brought Aquatabs (chlorine tablets) to treat water, and there are a couple small streams near us. We fill our bottles and I treat the water with the tabs. It tastes like chlorine, but at least the water is safe to drink.

After resting for about a half-hour, I feel considerably better, but still a little woozy. Pema insists on carrying my pack. With how I'm feeling, I don't argue with him as I think to myself, *This is what you get for making me hurry to the top of the pass!*

We continue on slowly as my stomach was still cramping, until we get to a stunning aqua-blue glacial lake that I believe is called Tilicho Lake. It's not very big and appears to be the remains of a melted glacier. Once past the lake, we begin to traverse some steep slopes, but luckily the trail is well cut into the side of the slope. Still, tripping here would not end well.

We stop for a spell at the Larke Rest House, a couple of long stone huts with blue metal roofs. The hut we enter has a long table straight down the middle. They serve basic food and drinks, and I think it's only open during the trekking season. I only have some apple slices, a granola bar, and water, as I don't want to upset my stomach any more. At least the cramps have subsided, and I feel that I'm now able to carry my pack again. Pema and I are the last of our group to stop here—we're running late, thanks to me.

Back on the trail about a half-hour later, we're high up on a hill see an area with short grasses and yaks grazing. In the valley a few hundred feet below us are some buildings, corrals, and herds of yak—it's the settlement of the local yaksmen and their families. A river runs through the valley, over some small waterfalls and the entire valley is lush with green grasses.

We climb over a small ridge, and the trail leads up a short, steep hill to the village of Samdo. This will be our last overnight stop before we continue on to the village of Samagaun at the foot of Mt Manaslu.

I'm so relieved to be in Samdo. It's 7:30 pm and already dark. I am exhausted, as it's been a long and trying day for me. We have rooms at the Tibetan Twin Hotel Lodge, which is basic yet comfortable. Everyone else has been here for a couple hours and they're in their rooms.

I see Jitesh and explain what happened to me. He's sympathetic, and I tell him I'll be okay after a good sleep. I go up the stairs to my room with a tea in hand, and unpack my things for the night. Pema comes to my room and asks me if I'm hungry. I tell him no thank you, I just want to sleep. I take some medication for my stomach, wrap myself up in two warm yak's-wool blankets, and fall to sleep.

Friday, September 8, 2017

I sleep right through the night and wake up at 5:30 am. My stomach is still feeling a little queasy, so I have a light breakfast of two hard boiled eggs and one chapati bread, plus a milk tea with some honey. I walk around the village a little, take some pictures, and make friends with some mules (I scratch them behind the ears, which they really enjoy). Then I just sit and relax in the wonderful morning sunshine.

Samdo at dawn as the warm Himalaya sun rises.

It's 9:00 am and we are back on the trail. Mingma Tenzi says it will take about two hours to reach Samagaun, where we will rest for a couple days before we climb to basecamp. This rest will be nice for all of us, as we have trekked many kilometres over this past week. Breakfast helped my stomach; it's feeling much better now.

From Samdo, we will be circumnavigating the valley around the base of Mt Manaslu into Samagaun. What a beautiful day! Just a few clouds and stunning views of the Himalayas and their snow capped peaks as we trek through the deep valley.

On the trail, we come across a stack of Buddhist prayer tablets, painstakingly hand-chiselled on flat stones by monks. It's eight feet wide, six feet tall, and about sixty feet long, with hundreds of these intricately inscribed tablets. Like the stupas and other holy sites and temples, it's only proper to pass these sacred sites on the left. This is good for blessings from Lord Buddha. It must have taken hundreds of years to compile all of these tablets—there are too many to count.

Next, we cross another small log bridge over a crystal-clear stream. The sound of the running water is soothing to my soul. Pema takes my picture on this bridge, then we continue on and pass some remnants of stone buildings. I wonder if they were here in 1956 when the first ascent of Mt Manaslu was made by the Japanese team of Toshio Imanishi and Gyalzen Norbu, with the help of twenty-one sherpas.

Not far from here I find a Nepali one-rupee coin on the trail. Mingma Tenzi tells me this is good luck!

CHAPTER 9:
HEAR TELL OF THE HEADLESS GOAT

We make great time to Samagaun—it only takes an hour to get here. We like to move it, move it! Three quarters of the way through the village, we reach the Tashi Delek Hotel and Lodge, where we will be staying for a couple nights.

I take my things upstairs to my room and laugh to myself. The stairs are quite tricky as each step is angled back towards the building. It's like they were originally built somewhere else and later transplanted here. You sure wouldn't want to navigate these babies after a few beverages! On the same level as my room in the corner of the hotel is a small room that actually has a Western-style sit toilet. Seeing this just makes my day.

Our arrival in Samagaun with the summit of Mt Manaslu
barely visible looming over the roof of the hotel.

I go down to the dining room and sit with Jite and Soham for lunch. I'm now feeling hungry again, so I have fresh chicken soup and a big plate of rice with green curry. I'm famished and it's so tasty, it takes me no time to eat it all.

Samagaun is very scenic, with a Buddhist monastery on top of a hill near the village's entrance. I can see it from here, its prayer flags fluttering in the gentle mountain breeze. When sitting and talking Jite, Soham and I decided to go visit this monastery tomorrow.

In the center of this wonderful village sits a stupa. It's about twelve by twelve feet across and twenty feet tall at its peak, white in colour with prayer flags strung off it's corners to other buildings, and you can walk through it. Inside on the walls and ceiling are old ornate paintings of Buddha.

All the hotels here are constructed from hand-masoned stone, concrete and skilfully crafted wood. Children play and chase each other about in the sunshine—lots of children, as well as many little white dogs so similar they look like clones. My guess is that the doggies here do lotsa cousin' lovin'.

Walking through the village, I hand out the rest of the Reese's chocolate bars and some candy to the children with their little rosy cheeks, runny noses and big welcoming smiles. The next thing I know, I have an entourage of children following me, with those little rosy cheeks all happy and smeared with chocolate. Pretty soon, they disperse and continue playing. I go back to the hotel to have a coffee and chat with Pema Sherpa about our days ahead.

Pema and I then go for a walk to the other side of the village. We stop to watch a helicopter fly in, land at the helipad, and drop off some supplies before flying off. We follow a trail through the village that hugs a creek as it meanders and winds it's way between rows of stone houses. As we walk we pass some women who are washing clothing in this creek.

The houses here all have two levels; the lower level is shorter, because it's for animals like goats, chickens, and buffalo. Unlike North American buffalo, these creatures are more like big Texas Longhorn cattle with jet-black hides. The ladder to the second level of the houses leads to the living quarters. This house design uses the body heat that rises from the animals to help keep the upper level warm during their brutally cold winters.

Small wood-burning stoves of clay are built into the wall or floor to cook, but with no chimneys, the houses are often filled with smoke. The chronic exposure to smoke causes many health problems and lung infections in the locals here, which often leads to death.

Back at the hotel, I have spaghetti and cheese sauce for dinner, then relax as we talk about the mountain and the days ahead in basecamp. I go into the kitchen for a spell and sit with the sherpas and porters. I don't understand Nepali, but by listening and watching their body language I can get the gist of some of what they are saying. The stove in here is also wood-fired, and despite an overhead hood and stovepipe, it's still somewhat smoky in here.

Speaking of smoke, that's enough for me. It's time to go to bed, so I leave for my room and settle in for the night.

Saturday, September 9, 2017

I hardly sleep a wink. Just as I fall asleep, a dog begins barking right outside my window. You wouldn't think that's possible, considering I'm on the second floor, but behind the hotel is a hill with a walking path about six feet from my window. *Ruff, ruff, ruff, ruff, ruff,* all bloody night long ringing in my ears, with an occasional howl.

You'd think somebody would have done something about this by now! A couple times I even get up, open my window, and yell at the dog to shut up. Like that's gonna do any good: this dog probably only understands Nepali or Tibetan, and not English.

Seriously though, we are supposed to be getting rest here and this is not working. As soon as it gets light outside, the little bugger shuts up, as though someone just flipped his switch off. Believe me, I had more explicit language at the time than you are reading here!

I came out of my room at 5:30 am and it's a gorgeous morning. The orange and red hues of the alpenglow off Manaslu from the sunrise feel spiritual. In the dining room, I have coffee and order some breakfast, fried eggs and chapati bread with raspberry jam. A few of the others come down for breakfast, and they aren't happy campers—they also haven't slept because of the dog .

The hotel owner is in the dining area working on the electrical bar that everyone uses to charge their phones. *If he cares about the clients at his hotel and his future business,* I think, *he would have done something about the dog last night.* He's within earshot. Maybe it's the wrong way to go about it, but I start complaining to Sergio and Lollo about the dog, just loud enough for the owner to hear. I sure hope this gets the point across, because we can't have another sleepless night. I didn't feel like talking with the hotel owner directly at this moment as I was a little angry with him for not doing anything to stop the endless barking.

Besides that, it's a beautiful morning. Jite, Soham, and I walk up to the monastery, but it's closed and no one is around. Up there, we see a few local girls carrying freshly quarried and masoned stones on their backs, and stacking them neatly in a large square pile. To carry the stones, the girls wear an L-shaped wooden rack on their back that is supported by a head strap. In the distance, I can see new buildings under construction. The stones must be for this purpose.

We turn around and head back towards the hotel. On our way, we see a woman outside the Samagaun Crafts Centre working a loom, weaving beautiful multicoloured Tibetan scarves. I ask her if it's okay to take a photo. She says yes, so I snap a quick picture.

Helicopters are coming and going, and more people are coming into the village now from the helipad. Back at the hotel, more expedition members have joined us, including some ladies from the International Women's Team: Samantha McMahon of Australia, Phunjo Lama from Nepal (although she also lives in New York), and Purnima Shrestha, a local woman from Gorkha, Nepal who now lives in Kathmandu.

Sergio, Jite, Lollo, Soham, and I are in the dining area, and I'm talking with Samantha and Phunjo as we hear even more helicopters landing at the helipad and flying over the hotel as they leave. Then a sherpa comes in with a white box, sets it on the table in front of the ladies, and pulls out what appears to be a black forest cake smothered in yummy-looking sweet icing.

The room goes silent as everyone's attention turns to this cake. Sam and Phunjo start to eat it, with everyone watching and probably wishing they could have just a taste. It was one of the funniest things I have seen in a long time. You know when you're eating something really good, and your dog is sitting there staring at you, smacking and licking its lips as it drools in anticipation? I found it very entertaining.

After this, I take in some sunshine up on the roof of the kitchen for about fifteen minutes, continuing to laugh to myself about the cake. I decide to wash some of my clothes, so I get my dirty laundry and take it to a sink mounted on the wall by the entrance into the courtyard.

As I'm finishing up with my clothes, the hotel's owner and a couple of his friends come into the courtyard as they lead black-and-white goat with a rope. It looks like an older goat, full grown with fair sized horns. They tie this goat to the fence, next to the small stone building that houses washrooms and a shower room. A few minutes later, as I'm wringing the water out of my clothes, the goat gets loose. As it runs past me, I try to grab it by the horns—but I miss and he makes his bid for freedom.

The hotel's owner goes after the goat and brings it back by the horns, this time making sure it's tied securely to the fence. I don't know yet that this goat is doomed, but I guess the goat itself sure knew it!

I hang my laundry in the sun to dry, then sit to take in some of the warm sunshine myself. Three children, around the age of three or four, come into the courtyard and start to taunt the goat. They are quite mean, poking and kicking the goat as it bleats in dislike. The children soon become bored with this and leave the yard.

Not long after, the hotel owner and his two friends return, carrying a small tarp, a couple knives, a machete, and a couple large pots. *Uh oh, this can't be good*, I think as they go over to the goat. They grab the goat: one guy holds it by the body and the other guy holds its head by the horns to stretch its neck out.

With one fell swoop and a ghastly thud, the hotel owner swings from overhead and plunges the machete blade three quarters of the way through the goat's neck. I hear a ghastly release of air from the goat's lungs, along with a sickening gurgling as its blood begins to gush out and run down its windpipe. The hotel owner takes another swipe with the blade and successfully creates a headless goat.

The guy who was holding the goat by the horns plops the head up on the stone fence. I can clearly see the goat's eyes still blinking and its tongue flicking in and out of its bloody mouth. Meanwhile, the other guy is holding tightly onto the headless body, which is still kicking and trying to run away. He picks the body up and tips it so that all its blood spills into one of the large pots. It doesn't take long to drain the corpse. He then sets the body on the tarp. The legs keep kicking for what seems like forever, but is probably only a minute.

After this, the three of them gut then skin the goat and spend the next two hours cleaning and prepping the meat to be eaten. They then clean the black-and-white hide and stretch it out on the tarp to dry in the sunlight. *I wonder what's for dinner tonight?* I think.

I try to take a shower, but the water heater isn't working properly and I don't like cold showers. I give up on this, too tired to worry about it. We will be going to basecamp tomorrow, where I know I can have a hot shower, so I'll wait until then.

I borrow Jite's pulse/oxygen finger monitor and place it on my forefinger. My resting pulse reads at 56 beats per minute (bpm) and my blood oxygen saturation is at 97%, which is actually very good considering we're at 3,780 metres or 12,400 feet elevation—almost the same elevation as Mt Robson's summit back home.

I have also been trying to get some word out to friends back at home that all is well. Supposedly there is Wi-Fi here, but it must not be working because I'm not getting any service. I was told that Seven Summits Adventures will have Wi-Fi available in basecamp, but I won't worry about it; I told everyone at home that once I leave Kathmandu, I'm essentially on the moon.

Before I left for Nepal, I went to the doctor to get medications for stomach upset, as well as antibiotics in case I picked up an infection, antivirals. I also have Diamox, a water pill that helps fend off acute mountain sickness and altitude illness, which are essentially the same condition.

The doctor said to take the Diamox before climbing to basecamp. I ask one of the sherpas about it, since he has more experience in the high mountains. He says that I shouldn't take it unless I have symptoms of acute mountain sickness. I mention this to Jite, who says he was now taking Diamox and that I should, too. I think I will do as the sherpa advises, as I feel he is more experienced in these matters.

Sunday, September 10, 2017

Again last night, no sooner had I fallen to sleep, that damn dog was at it again. I yell at it to shut up a couple times, which does no good. I try wrapping my head in the covers and even stuff some toilet paper in my ears, but I can still hear the little bugger.

My heart sinks when I hear a whack and the dog yipping, like someone just kicked it. I guess the hotel owner got my message when he heard us talking at the table, but I never wanted him to hurt the dog. All I thought he might do was tie up the dog away from the hotel.

Now that the dog is quiet, I notice a noise in the wall behind my head, like something is scurrying up and down behind the plywood inside the wall. It grows louder. *Oh my God, I can't win,* I think.

I get dressed, grab my headlamp, go outside up the trail on the hill behind the hotel. About fifteen feet away from my window I stop, kneel down, and wait. It doesn't take long. Something runs up the wall and enters a hole near the roof. Seeing more of these shapes running up the wall, I aim my headlamp towards them and switch it on.

RATS! The poor dog was just doing what comes naturally—trying to get rid of the rats! There's nothing I can do except change rooms. I go back to my room, grab a couple things, find a vacant room, and lie down to sleep. Finally, I get a decent sleep.

Today we will be climbing to basecamp at 15,400 feet (4,700 metres). Our elevation gain will be about 3,000 feet (914 metres), which is huge for a single day. For proper acclimatization to avoid altitude illness, you're only supposed to gain about 1,000 feet (305 metres) per day for three days, following with a day of rest before you continue. There's nowhere to camp between Samagaun and basecamp, though, so our only choice is to keep going. This makes me wonder how many people arrive at basecamp with symptoms of altitude illness.

I hear lots going on outside so I open my door to have a look. Many locals are gathering in the courtyard, preparing to work as porters to carry all the supplies for our basecamp up the mountain. They are paid according to weight, so the sherpas use a scale similar to a fish scale to hang and weigh everything. I have heard they earn $35 USD for a basic load from here to basecamp.

Many of these local porters are young women from the village, but our personal gear bags will be carried by our expedition porters, the young men I've mentioned previously. It's crazy here with all the porters and chatter going on, It's all very loud and entertaining.

Local porters preparing to haul our supplies up to basecamp.

With the entertainment of the porters gone now as they make their way up the mountain, I go into the dining room and sit with Sergio, Lollo, Jite, and Soham. We are all so happy that the dog was not barking all night and we got some sleep. Sergio laughs and says he heard me go out and kick the dog.

"That wasn't me!" I tell him in shock. I am a dog lover and would never hurt it. The only thing I would do is get a rope, take the dog away from the hotel, tie it to a tree for the night, then release it in the morning.

I ask Sergio if he remembers when I spoke loudly about the barking dog within earshot of the hotel owner near us. Sergio says yes, and I say that my plan obviously worked. The hotel owner must have heard me and finally done something about it—just not what I thought or wanted him to do to the dog.

I tell the guys about the rats I saw and that the dog was actually barking for a valid reason, and now I feel bad for the dog. They laugh.

After a nice breakfast with milk tea, we get ready to climb to basecamp. We leave shortly after 8:00 am. Jite and Soham are the first to leave, and I follow them. Pema is behind me with Purnima Shrestha behind him. I learn that Purnima is a photojournalist who works at a paper in Kathmandu called the *Karobar Daily*, and this will be the first mountain she has ever climbed. She plans to climb Mt Everest next spring. *Wow, she is very ambitious*, I think. Being a photojournalist also explains why she has such a nice camera with her, which I quite admire as my father was a hobby photographer.

CHAPTER 10:
BASECAMP AND BEYOND

We're moving at a good pace as we leave the village. We come across a few beasts of burden, pack yaks coming from Samdo headed towards Samagaun. I stop to take a picture of one of the yaks, then catch up with Jite and Soham. Pema slows down and walks with Purnima, so the guys and I get ahead of them. We turn off the main trail and go uphill on a smaller trail. The trail gradually gains elevation, and we stop near a creek in the shade for a break. I also decide to take off a layer or two, as it's beginning to get pretty warm.

We climb a steep drainage path through the knurled, knotted trees. It's nice and cool climbing in the shade of these trees, but it doesn't last long. Soon we're above the treeline and the slope is getting pretty steep, the trail switching back and forth to ease the climb.

From here, I can see Samagaun behind us, as well as Birendra Lake below us. Its emerald green glacial waters shimmer and sparkle in the bright midday sun.

Looking back towards Samagaun and Birendra Lake.

Many trekkers on the trail are going up to basecamp and no further. A few of these are older retired couples. Jite and Soham have gotten much farther ahead, but I'm not alone as Samantha is right behind me. I ask her if she would like to get past me, and she said no, my pace was good.

Sammy and I chat a little as we climbed. The trail is among shrubs and beautiful multicoloured flowers with many bees visiting them for their nectar. In a couple spots, the flowers are so plentiful I can smell their sweet fragrance.

This immediately brought back memories of being in Jasper as a boy with my mom and dad, with the smell of sweet mountain pine and spring flowers. I then have the powerful sense that my folks are with me in spirit and actually experiencing all of this through me. I continue on, feeling immense pride.

Pema has caught up to me now and passes me on a side trail. Pema tells me to move over to let Samantha pass. Samantha said that she was okay behind me and wanted to stay there. About ten minutes later I stop to take a few pictures, and now Sam continues past me.

A short while later, we come to a temporary teahouse that is only open during Manaslu's climbing season. This is about halfway to basecamp and everyone on the trail is stopping here for a rest.

At this elevation, I can't believe it when I see a couple Chinese guys smoking cigarettes—and they're not even sherpas. Nothing against smokers, as I smoked for thirty years and quit about nine years ago—I'm just surprised that they, foreigners like myself, are able to climb at this altitude and still smoke. I couldn't do it! They must be very young.

After resting about a half-hour, I continue climbing with Pema alongside. Soon we need to cross a fast-moving glacial stream that is bounding down the slope and over a couple small waterfalls. As usual, I like hearing the sound of the water as we get closer. Up ahead, yaks headed up to basecamp burdened with supplies strapped to them are crossing the stream. Then it's our turn. The rocks are wet and look slippery. I make the crossing and it wasn't too bad, especially given my past history with crossing streams.

We climb out of the shrubs and flowers. The terrain becomes rockier, with moss and grasses as well as tiny alpine flowers. The yaks have now taken a wider trail with a gentler grade that goes off to the right. We cross another fast-flowing stream and continue on up the trail.

Ahead I see a group of people sitting in the grass and moss of a somewhat flat spot in the slope. When I get there, a sherpa asks me if I would like to rest and have some tea and cookies. I answer yes. (Who can pass up cookies and tea? Not me!) The sherpa introduces himself as Lakpa Sherpa, the managing director of Pioneer Adventures. "I guess Lakpa is a common name in Nepal! This Lakpa is different than the Lakpa Thendu Sherpa of Adventure 14 Peaks I met in Kathmandu. This Lakpa is heading up the International Women's Team on Manaslu this year. Sitting with him are a couple of his sherpas and Phunjo Lama, who I met in Samagaun with Samantha.

It's very cool that Lakpa had one of his sherpas from basecamp bring down a big stainless-steel thermos of hot tea and cookies. At this point, we are three quarters of the way to basecamp and I'm feeling pretty good. No headache, which is usually the first sign of altitude illness.

We now start to pass the local porter girls, who are descending after bringing our supplies up to basecamp. Every time they pass, I tell them *"Dhan'yavad"* for all their hard work. The girls who don't seem to understand me look puzzled, so Pema tells them in what I believe is Tibetan what I said.

Then they giggle and give me a big smile, their perfect white teeth beautiful against their sun-brazened skin and rosy cheeks. Pema seems to get a kick out of this, and he laughs along with the girls.

We climb on, and I can now see some of the glacier beside us that comes down off Nadi Chuli Peak and Mt Manaslu, feeding Birendra Lake below us. We come over the final ridge and see basecamp emerging, as well as the east pinnacle of Manaslu glowing magnificently in the bright sunshine.

Pema goes ahead of me as I take a couple pictures, I keep moving and see a city of mostly yellow and orange tents scattered everywhere amongst the rock and moraine. Basecamp is where the Manaslu glacier once was a long time ago, now this is a glacial moraine field with steep rocky vertical slopes on either side that rise to become icy peaks. There are also a couple streams that meander through base camp of which one of these begins its journey through our tent city by cascading down a rockface.

Lower basecamp on Mt Manaslu at 4,800 metres (15,750 feet).

As I continue up the trail through basecamp I see tents everywhere assembled in many groups according to their country or expedition company. The first group I see is beside the helipad, overlooking the valley and Samagaun below. This is the Chinese camp, with a Republic of China flag proudly flying in the breeze. To my right, I see the Swiss flag, and much higher up is the Russian flag.

My camp is the Satori camp (Satori is the mother company of Adventure 14 Peaks). Everyone else in the camp has booked through Satori; I am the only one here with Adventure 14 Peaks. We each get a two-man tent to ourselves, to house us and our gear. Our large dining/common tent, about twenty feet long and ten feet wide, has enough room for two long tables set end to end, with eight chairs on either side. The kitchen has its own tent, plus another tent for kitchen storage. There's also a small shower tent, about four by four feet, as well as pee and poo tents of the same size. (The poo tent, of course, is brown!)

The pee tent has some designated rocks inside on the ground that you pee on. The poo tent is another beast altogether. The camp helpers line a big plastic barrel (about three feet tall and two feet across) with a heavy-duty plastic bag or two. They set the barrel on the side of a drop-off, and build up two sides to the top of the barrel with flat stones. Essentially, you get a stone platform that enables you to squat over the barrel. One side is left open so they can pull the barrel out and change barrel when needed. When 1/2 to 3/4 full a porter takes the barrel strapped to his back down the mountain to a disposal site.

They don't want any liquid waste in the barrel. I am thinking because of the sloshing around during transport causing a very hazardous out of balance situation for the porter on the narrow steep trail. This is where the poo-tent rule comes into play: while making a deposit into the barrel, you cannot pee in it. You must pee on the rocks on the outside of the barrel as you poo inside the barrel. As a guy, this isn't too difficult, but I have no idea how the ladies deal with this—and I'm not sure that I want to know either. Maybe yoga would help with this.

I'm told there are 350 to 400 people in basecamp this year, and it has been getting busier ever since the cost of the climbing permits for Cho Oyu peak in China have increased substantially. The Manaslu circuit trek has only been open to foreigners for a couple years now, and it's part of the Great

Himalayan Trail as well as an ancient salt trading route. As far as trekking routes go, Manaslu is the gem of the Himalayas and becoming popular with trekkers from around the world. The route travels through ancient Himalayan culture and virtually untouched remote areas, as we have experienced on our trek to Samagaun.

I unpack and settle into my tent, which is right next to the common tent. My laundry from Samagaun isn't completely dry, so I hang these items up again on a tent fly cord tied to the common tent. I then get some stones and use them to make a nice flat entry into my tent.

As I'm moving the heavy stones, I feel the effects of the altitude and need to catch my breath often. Working bent over with the heavy stones probably isn't helping much either, as this position isn't great for taking deep breaths.

Beside my tent in basecamp with my laundry drying.

After improving access to my tent, I join Jite, Soham, and Sergio in the common tent, and meet Ina Condrea of Romania. If successful, she will be the first Romanian woman to climb Mt Manaslu. She actually lives in Switzerland, though, as a language teacher. She is also an ultramarathon, mountain trail runner.

Ina is a little concerned that her gear bag with her warm clothing hasn't reached basecamp yet, and she's feeling a little chilly. I get my new Rab down-filled belay parka from my tent and give it to Ina to wear until her bag shows up. Funny, as I hadn't even worn it yet, but my father did raise me as a gentleman! So far in camp it's Ina, Jite, Soham, Sergio, Lollo, and me, and the six of us end up becoming like family. (To her relief, Ina's bags make it to camp about an hour later.)

Our common tent has a heater, a charging station, and all sorts of snacks as well as juice mix, coffee, tea, and hot chocolate mix. Soham gives me his pulse/oxygen monitor and tells me I can hang on to it to use throughout the expedition as Jite has one too and he can use Jite's. Speaking of snacks, we just had dinner: yak meat. It was very good, well cooked and tender.

I'm very tired, so I'm going to go and crawl into my sleeping bag for the night, *subha ratri*!

Monday, September 11, 2017

I didn't sleep well last night, as it took me a long time to fall to sleep and I woke up a few times. Upon waking this morning, my pulse is 92 bpm and my oxygen is at 77%, which is low and might explain my lack of sleep. I was planning to trek up to the crampon point for some training and acclimatization, but given my lack of rest, I'm not sure about doing this. I will see how I feel after breakfast.

The crampon point is where the foot of the Manaslu glacier begins. To go any farther, we need to don our crampons as it's all ice and snow past this point. It's also where the fixed lines begin and carry on up to the summit. Manaslu's crampon point is at 16,728 feet (5,100 metres) of elevation.

I'm feeling better after breakfast, so I put on my Scarpa climbing boots. These are the same boots I wear while ice climbing back at home. I also pack some warmer clothing, snacks, water, and begin my trek. Wow—there are so many tents and people here, it's like its own international city! No matter the race, colour, creed, or religion, everyone here is very friendly and helpful towards each other.

It's a nice day today with a few scattered clouds. The Himalayan sunshine feels so intense at this altitude, and I have never experienced the likes of the vistas of the surrounding mountains. Especially the very impressive Samdo Peak directly across the valley from us. Manaslu's white sunlit mass against the dark blue sky can only be viewed with good sunglasses on. Otherwise, the sun and its reflection off the snow is too intense and could cause snow blindness.

I find as I gain altitude that I'm doing pretty well. If I start to feel out of breath or my heart pounding in my head, I slow down and take in more air. This strategy seems to be working quite well for me. I just need to concentrate on pacing myself according to my oxygen requirements.

Once I climb a little ways above upper basecamp, I come to a big boulder that has a metal memorial plaque attached to it. It reads: *In memory of Jafar Naseri, Iranian mountaineer on the top of Manaslu, 10 May 2012*. It also has his picture above the inscription. I stop, pay my respects—*Godspeed to you, brother*—and continue on my way.

Since 1971, there have been eighty deaths on Manaslu. In September 2012, eleven lives were lost in a single avalanche, with the mass of the slide completely wiping out Camp Three. Mt Manaslu as well as Mt Annapurna have been nicknamed the killer mountains, Manaslu due to its warmer, wetter climate causing many large avalanches.

I continue up to where there are fixed ropes that take you up some slippery wet rock steps, where there's high exposure to a deadly fall. Since I'm alone and no one else is around, for a safety sake, I stop here, rest a bit, then turn around and head back down. I get back to camp and I'm quite pleased with how it went, as I had no problems or adverse symptoms from the altitude.

When I arrive back, I hear that we could possibly be climbing to touch Upper Camp One tomorrow at 18,700 feet (5,700 metres), then come back to basecamp. This climb will help with the acclimatization process. The purpose is to get your body to adapt to the lack of oxygen by producing more oxygen-carrying red blood cells, making you more efficient with the oxygen available.

At over 26,240 feet (8,000 metres), there is only one third of the oxygen available for your cells at sea level. This is mainly due to the low atmospheric pressure at these heights, essentially making the air thinner. Per cubic metre of air, the percentage of oxygen and nitrogen is the same (about 20% oxygen to 70% nitrogen), but there is simply less of both gases, as there is less pressure to pack them in, so to speak.

Under 10,000 feet (3,048 metres), the pressure actually helps you breathe by pushing the air into your lungs' alveoli. Every time you exhale, you create a vacuum in your lungs. Above this altitude, the reduced pressure also reduces this vacuum, meaning you actually need to work to draw air into your lungs. The higher you ascend, the harder you need to work at just breathing.

Wearing and using supplemental oxygen only alleviates this somewhat, as it feeds you more oxygen per cubic metre than you would get from the atmosphere. You can still suffer hypoxia and altitude illness if you spend too much time over 26,000 feet (8,000 metres). It's known as the death zone because there is not enough available oxygen here to support life, because of the lack of pressure it is very difficult for your cells to take in oxygen and you are literally and slowly dying. The only way to stop this would be to wear a pressure suit with oxygen, like the astronauts do, or descend.

Back to acclimatizing! It takes time for the body to build enough extra red blood cells to deal with the lack of oxygen. For mountains that are just around 8,000 metres, climbers will spend about twenty days to a month on the mountain to give their bodies time to adapt. On Mt Everest, at 8,848 metres, many climbers will spend two months on the mountain so their bodies have enough time to build up enough red blood cells. As you climb higher over 8,000 metres, the pressure decreases exponentially, making it even more difficult for the reduced oxygen to make it to your starving cells.

For example: to assist in adapting on Manaslu, we climb to basecamp, then rest the night. After a day or two, we climb to Upper Camp One, then back down basecamp to rest for another day or two. Then we climb up to Camp Two and back down to Upper Camp One to sleep for the night, then we go back down to basecamp and rest there for a few days to just over a week waiting on a clear weather window forecast to go to the summit.

Once the weather has cleared up and we've received a good forecast for the summit. We then plan our ascent to make the summit on the morning with the best conditions. So we leave basecamp to climb to Upper Camp One and sleep the night. Climb to Camp Two and sleep the night. Climb to Camp Three and sleep the night. Climb to Camp Four and sleep for three hours. Then, wake up between 11pm to midnight, have a bite to eat. After this we put on our down suits, boots, crampons and oxygen masks then begin our climb to the summit, also called the summit push! After some time on the summit to take pictures, we descend to Camp Two and sleep the night, descend to basecamp and sleep the night, pack up our gear and then descend to Samagaun. This example can vary a day or two either way, depending on the climber's condition and strength, and (of course) the weather.

Sometimes you can spend a week or two stuck in basecamp in cold, wet conditions, waiting for the weather to clear or avalanche hazards to subside. For the summit push, you want clear weather for obvious reasons.

Bad weather can be deadly. The 1996 Mt Everest tragedy occurred when an unexpected monsoon swept up into the mountain. The movie *Everest* and the book *Into Thin Air* by Jon Krakauer are based around this tragedy.

Even on a clear day, high winds can be dangerous. It takes energy to fight strong winds, and reserving your energy is crucial—some climbers have died just from exhaustion. Also, your body needs oxygen to stay warm, and with the lack of it you can easily get severe frostbite on your extremities. The wind-chill makes climbers much more prone to this.

I'm hoping to have one more day in basecamp before going to Upper Camp One, as I haven't yet pulled out my high-altitude boots and fitted them to my feet with the summit socks on. My summit socks are a nice new pair of thick wool socks that only get worn on summit day so they retain their loft and warmth. The high-altitude boots have a semi rigid outer shell with lacing and Velcro to stay secure. An insulated inner boot also laces tight and locks in place with Velcro.

Many climbers make the mistake of getting boots that fit snugly, which puts them at risk of losing toes to frostbite. You need wiggle room for your toes, It also helps to try and scrunch your toes as you step to keep them moving thus increasing blood flow. At high altitudes, your feet swell, reducing the circulation as your toes squish together. Have you ever noticed when you are flying and take your shoes off, they're so much tighter when you try to put them back on? Airplanes keep their pressure at about 8,500 feet in altitude, and your feet swell enough to make your shoes tight. Imagine how much they would swell at 20,000 feet and above!

I haven't climbed or walked much in the big high-altitude boots, and want to take them out for a spin before we actually begin the ascent. I can get accustomed to them and make sure they are perfectly adjusted so I won't have to mess with them once on my way.

Today with Mingma Tenzi's help, I set up my harness for climbing fixed ropes with an ascender (or jumar), also called jumaring or jugging. The ascender is a handle with a rope grab built in to it that clips on the fixed rope. It will only slide up the rope but not down, so you can use it to pull yourself up the rope as you climb. It's also for safety and will stop you if you fall. Tied to my harness are three slings: one to my ascender, and two with carabiners to tie off to the rope as a safety in case I fall. This will be the first time I have jumared or jugged up fixed ropes.

Until now, I have always climbed traditional style on mountain ascents (or "trad"), which means placing your own anchors and being belayed as you climb. The bottom climber is stationary and anchored to the ice or rock, and belays the lead climber, who places anchors as they climb and clips the rope into these anchors. If the lead climber falls, the partner below stops the fall through the anchors, rope, and the belay device attached to their harness which is anchored to the rock or ice.

Lead climbing can be quite dangerous, especially if the leader is ten or more feet above the last anchor they clipped the rope into. The leader will fall the ten feet to the anchor then ten feet below the anchor. Because of the slack in the rope this will result in a fall distance of twenty feet or more. Now near the end of the rope, the lead climber anchors himself to the rock or ice, sets up a belay anchor and belays the bottom climber up, who collects the anchors on the way. Then they repeat this sequence, known as a pitch, to continue up.

Back in the common tent, everyone else is playing cards. I'm trying to get caught up with writing in my journal. I'm also looking over the climbing permit, and it turns out that I am the oldest climber on our team thus far.

Ina has also gotten us all addicted to Penotti hazelnut white and dark chocolate spread. We don't spread it on anything, though: we just eat it out of the jar with spoons. It's absolutely delicious and we have dubbed it Himalayan superfood.

After catching up on my journal and satisfying my craving for sweets, I finally shower. The shower tent setup is simple—it's just a bucket with a tap on it that you hang in the top of the small tent. It's more of a trickle than a spray, but the hot water feels great!

Most people in camp are also doing what we call steam baths: holding your head over a large stainless bowl of steaming water with a few drops of Tiger Balm oil in it, draping a towel over your head to trap the vapours as you inhale them. I guess it works wonders if you are congested or want to open up your airways. I have no interest in trying this now as my airways feel clear and even though we are as family here. I do not want to get too cozy and complacent as I have many things to concentrate on.

CHAPTER 11:
HIMALAYAN HEATSTROKE

I find it hard to believe that I have already been in Nepal for two weeks. It's gone by so quickly. I find that sometimes I'm lonely and bored in basecamp, as I am the only one here from Canada who speaks primarily English at home. With everyone chatting away in their native tongue, I sometimes feel left out. I have never been much of a social butterfly anyways.

Maybe I'm feeling a little homesick, or maybe it's the altitude. I am also much older than every one here. I've realized that everyone here has someone waiting for them to come home safely. I have no one waiting for me at home. The benefit to this is that if something happens and I don't make it out of here, then I will not have caused anyone else's heartbreak and grief. *Okay, that makes me feel better,* I think, telling myself, *Suck it up buttercup!*

I walk around basecamp and try to talk to people about climbing, but I find that many people here are not climbers, as in rock climbing, ice climbing, or even alpinism. A couple of people even said to me that the style of climbing here on Manaslu is true climbing, and everything else is essentially child's play. Interesting—by the sounds of it, I don't think they've ever tried mixed climbing or even climbing up a sheer rock face.

There are, however, a lot of marathon runners up here. I never would have thought of them climbing dangerous mountains, but I guess if anyone would have more success with the altitude, it would be a marathoner. After all, there are no difficult technical sections on this mountain; it's basically a really tough hike to the top of a really big mountain.

I guess I'm out of my element, with not much hiking experience compared to my time spent climbing sheer or overhung rock and ice with technical crampons and ice axes. It's like comparing apples to oranges. There are many different styles of climbing—none is more difficult or better than the other, it's just a matter of personal opinion. I just need to remember that I'm here to train and prepare for my goal of climbing Mt Everest.

Canadian Rockies Youth Society flag for the summit.

Before I left home, I had a flag made up for Canadian Rockies Youth Society to fly on the summit. It's about two feet by three feet with a royal blue background; in the middle is a white shield bordered in red with mountains in the middle of the shield and a banner underneath, both in red. Across the top of the shield, above the mountains, it reads "Canadian Rockies" in red, and "Youth Society" in white across the banner.

I set it up on the supplies tent and take a couple pictures. It's going to look awesome at the summit! I also plan to fly this flag proudly from the summit of Mt Everest.

Well, time to turn in for the night. *Subha ratri.*

Tuesday, September 12, 2017

I had a good sleep last night. My pulse/oxygen levels this morning are 79 bpm and 77%. I'm a little concerned, as my pulse should be lower and my oxygen levels should be higher.

When I wake up, the tent is damp. We're socked in with clouds, bringing the humidity way up; it feels as though it wants to rain. The air is so saturated that you can see mist just hanging in the air around you like you're standing near a waterfall. We're supposed to go to Upper Camp One today, but plans might change if it's raining or snowing. I know I sure don't want to get wet and cold. With this humidity, rain wear won't help much—it's the air itself that's wet.

I get my things packed and ready to go anyway, then head to the common tent. Once I'm inside, the rain starts to fall. Everyone in the tent agreed that we likely weren't going anywhere in these conditions.

Just before breakfast, Mingma Temba came in to tell us what we were suspecting: we would stay put for an hour and see if the weather changes. He said if things improve, then we could go up to the crampon point and put our crampons, harnesses, and helmets in the sealed plastic barrels there. Then we won't need to carry them there when we leave for Upper Camp One.

We had breakfast and it's 8:30 am now. The rain has changed to heavy wet snow. Outside it's slippery and the visibility is poor due to the snow. I'm thinking it's definitely not the day to go to Upper Camp One because of the added avalanche hazards on the glacier, which isn't that far from here.

We then do what we can to pass the time. A couple people go to their tents and sleep, and others play cards. I write in my journal as I have never been much of a card player. We have lunch and things haven't changed much; now it's a rain-snow mix. At 3:30 pm, the rain and snow finally stop, and unbelievably, the sky opens up to complete cloudless blue. Now we can at least trek to the crampon point.

* * *

We're back from the crampon point now and it went well. I wore the high-altitude boots and I love them! Considering the size of them, they are light and comfortable. Their grip is surprisingly good, even on wet or icy rock, but explained by their soft, almost foamy Vibram souls. I paced myself well while climbing and had no issues with the altitude. Pema and I made it to the crampon point in just over an hour, as he predicted. We left some gear in the plastic barrel then sealed and locked it. I'm pleased with my performance and feeling confident.

Back in basecamp, two more members have joined us: Riccardo Bergamini and Alessandro Corazza, both from Italy. Having worked with Italians in the past, I find it entertaining when you bring them together. Even if they're not related, it's like a family reunion: loud and boisterous, with arms flailing all over the place.

Oh, lo and behold, the clouds and bad weather are closing in on us again! At least the gods opened the sky long enough for us to take our legs out for a trek today. When we come back, the cook and his two helpers have made us a "Welcome to Manaslu" cake and decorated the common tent with garlands—a nice touch. The cake was sweet and pretty good.

Inside of our common/dining tent with garland hanging.

Wednesday, September 13, 2017

I had a decent sleep last night and this morning my pulse/oxygen levels have improved, 92 bpm and 83% oxygen. Last night was the first night in the tent that I didn't feel like I was lacking oxygen. I slept well even though I was awakened by a heavy downpour that turned to snow again.

When I wake up, the snow is still coming down pretty heavily. I get the feeling today will be a slow start for anything if at all. I hear avalanches all around us constantly—some so loud and close that I've had to look out of the tent just to make sure none of them were coming our way.

It's just after breakfast now and the snow has changed back to a rain-snow mix. Because of the rain, much of the snow we've accumulated has melted.

Right after breakfast, we had our puja (a Tibetan ceremony for blessings and safe passage on the mountain). Lama Guru, the senior monk at the Buddhist monastery in Samagaun, at seventy-seven years old climbs to basecamp and performs the ceremony for us. In camp we have stone altar about four feet square and about four feet tall. It has a fire box built into its side, where juniper is set to smoulder. The smoke carries the prayers to the gods on the breeze during the ceremony.

A four-foot wooden pole extends from the top of the altar, from the top of this are strings of Tibetan prayer flags that are strung out in different directions and at their ends attached to tents or rock. Lama Guru has prepared small statues (made out of dough, I think) depicting Buddha, Manaslu itself, and various deities we will pay homage to.

On the altar is a bowl of rice, a bowl of flour, candy, soft drinks, pastries, oils, wine, and a bottle of whisky—offerings to the gods that we will also partake in after the ceremony. Before the puja, we place any climbing gear we want blessed (boots, axes, etc.) at the back and base of the altar. During the two-hour puja, Lama Guru and his assistant chant Buddhist prayers in unison for our blessings. He occasionally rings a hand bell and his assistant beats a hypnotic rhythm on a drum. They are asking for our safe passage on the mountain.

During the puja, light sleet continues to fall so we all need our umbrellas, holding one over Lama Guru as he chants the prayers for our safe passage. We're also drinking a strong homebrewed rice alcohol called sherpa soup. It tastes a little piney, almost like juniper, but after a couple sips that goes away and the warm fuzzy feeling creeps in.

Before we finish, we rub some flour from the altar onto one of our cheeks for a blessing. We throw the remaining flour from our hands up over the altar and yell out in unison, "Oooooohhhaa!"

After the puja, Lama Guru places twisted strings of yellow and red around our necks and ties them as a necklace for our personal blessing. Now we all indulge in the treats on the altar, and the whisky as well. Needless to say, at this point we are all feeling pretty good!

We gather some money and give this to Lama Guru as a donation for his services. My guess is he takes away a couple hundred dollars, probably more with larger groups, from each puja he performs. I believe that he stays in basecamp for a couple days just performing ceremonies. Of course, the sky decides to clear after our ceremony. The gods must be blessing us!

It's just after 3:00 pm. The clear sky didn't last long; since lunch, it's been raining, drizzling, raining, drizzling, nonstop. It would sure be nice to get into those high-altitude boots and do a little more climbing to keep the legs strong.

The last three members of the group have joined us: Mathew Eakin from Australia, Nadev Ben Yehuda from Israel, and last but not least, Bobi Zvezdata from Bulgaria. There were three others on the permit, but they won't be coming (I'm not sure why). So now the Satori Expedition family is complete!

I sure hope this weather turns around for us soon. Sitting in a cool damp tent all day really sucks, but this is part of the Himalayan game. Hurry up and wait for weather, sometimes for a week or two, then suddenly—*go, go, go!* The good thing is that this is better for acclimatizing, unless you come pre-acclimatized from hypoxic training or from another mountain. In those situations, extra time at basecamp is detrimental. Your body is already used to being at much higher elevations, and you begin to lose the extra red blood cells that you worked so hard to build up.

Thursday, September 14, 2017

I had another good sleep last night and my waking pulse/oxygen is 69 bpm and 88%. Great, it seems I'm acclimatizing just fine.

It's a nice morning when I come out of my tent, only a few clouds and sunshine with a warmer temperature, or so it feels. I hope it stays nice all day.

This morning, I see some inexperienced guys trying out their boots for the first time. They look like drunk sailors, literally tripping over their own feet. *Wait until you put on the crampons,* I think, amused. *I mean, wait until your sherpa puts your crampons on for you.*

After breakfast, I decide to take my legs out for a trek to the crampon point. The trek is going well; it's easier than before, and I feel like I have more energy now.

There are others out today stretching their legs too. I decide to climb and cross over the rock with the water flowing over it to reach crampon point at the foot of the glacier. The views today are spectacular! I see a couple of good sized avalanches on Nadi Chuli Peak, and the east pinnacle of Manaslu is gleaming in the sunshine. I made good time, too, as it only took an hour and ten minutes to get to the crampon point. My throat was a little sore when I got to the point; the air is so dry up here that it draws moisture from you. I drink a little more water and my throat feels fine. I spend a couple minutes here and then head back down.

I arrive back at basecamp feeling good. I have a bit of cough from the dry air, but with a little water it dissipates, just like my sore throat. I was pretty hungry for lunch—spicy chorizo sausages, pasta with sauce, and chapati bread.

After lunch, I spend the rest of the afternoon just relaxing, enjoying the warmth from the Himalayan sun and the spectacular views surrounding me.

Staying in big five-thousand-person camps while working in the energy industry has made me into a bit of a germaphobe, so I notice immediately Mathew has picked up a cough that doesn't sound like it's from the dry air. I ask Matt about it and he says it's a throat infection. My intuition was correct. *Uh oh,* I think, because we are all so close to each other in a confined space, the common tent.

Then Jite comes into the tent and he has the same cough. Now I'm worried! Right away, I go to my tent and take one of the antivirals I brought from home.

After dinner, I write in my journal. The sherpas come and tell us if the weather is good tomorrow, we will go up and touch Upper Camp One. Finally! After hearing this, I go relax in my tent and fall asleep for the night.

Friday, September 15, 2017

Last night my sleep was off and on. I woke up a few times, but I feel not too bad this morning. My pulse/oxygen is even better this morning, at 61 bpm and 92%.

I'm usually the first person up, so I go to the chilly common tent and fire up the heater. Ina and Jite are usually the next people in the tent and this morning is no different. We're thinking today will be a good day, and indeed, the sherpas tell us the forecast is good and we will be heading out for Upper Camp One shortly after breakfast. I make sure to put on sunscreen before we head out.

We make good time to crampon point and I feel great. Some clouds move in, but it's nothing to worry about. It's just a little hazy, which is good while on the glacier—glacial travel in the sun can be as hot as a desert. The sun's rays here fry you fast, as there is much less atmosphere to block the UV rays coupled with the reflection off the snow. As a matter of fact, many people get sunburn under their chins and noses from this intense reflection. So I'm actually happy for the haze.

Finally, I get to use my crampons on Manaslu's glacier. I put them on my boots and we take a few minutes to eat a snack and drink some water. As we begin climbing the first 100-200 metres up, the toe of the glacier is fairly steep, at about 45°. This angle reduces to about 15-20° as we get on top of the main body of the glacier.

This is now a vast expanse of crevasses. Most can be jumped over, and the trail weaves around and between the larger ones. There are fixed ropes to clip into, but a few of the aluminum pickets that anchor the rope have come out and are just lying on top of the snow. Even while you are clipped into the rope, you would fall a long way into a crevasse before all the slack was taken up thus arresting your fall.

Two of these pickets I stomp back into the snow and then pack some snow on top of them so the sun doesn't melt them out, it's amazing what a few well-placed pickets can do and how much weight they can actually hold. We get behind a couple climbers with no glacier or crevasse experience at all it seems. I see them stepping right on the edge of the crevasse before they jump which is a really bad idea. One of them found out when the edge gave way and he almost fell into it as his leg dropped out from under him. I thought to myself, that almost became a rescue.

As I learned from Jay Mills in my glacial travels and crevasse rescue course: before stepping near the edge, use your trekking pole or ice axe to poke or prod the edge to make sure it is in fact solid! We continue up through the crevasse field, stopping for a rest at a spot stomped into the snow where about eight to ten others are taking a break. Pema chats with a couple of his sherpa brothers as I drink water and eat a snack.

It has been very warm since we left basecamp, so I'm only wearing softshell pants and a T-shirt. I'll need more sunscreen, because even with cloud cover, you can easily burn up here. I look everywhere in my pack for my sunscreen, but I can't find it. I must have left it on the table in the common tent. Darn!

Time to get going again; I can borrow some sunscreen when we get to Upper Camp One.

About two hours later, I can see Lower Camp One and where Upper Camp One is up on the shoulder above the glacier. It looks to be only about another hour and a bit to get there. The clouds have left and the sun is blazing. I can feel it frying me—it's incredible how intense and hot it can be, especially with the reflection off the snow.

I left my umbrella back in basecamp because I didn't think I needed it, but I wish I had it now. The only jacket I brought was my black Arc'teryx—a bad colour choice, as we all know what black is like in the sun. I would totally cook in it, and my sweat wouldn't evaporate either.

A horrible yet familiar feeling overcomes me, similar to Larke Pass but different. This time, I think it's heat-related. I stop sweating and my skin is hot to touch. I also have a sudden, pounding headache and I'm feeling very weak. I'm left with no choice but to get off the hardpack trail and sit on my pack in the deep snow.

I've had heatstroke in the past and that's exactly what this feels like, except much worse. Heatstroke is a medical emergency and can become fatal if it isn't taken care of.

I know at this point that I'm not going to finish the climb to Upper Camp One, and that I need to be back at basecamp to recover. I begin to come out of it after sitting a half-hour, having consumed some cookies, a melted Snickers bar, and lots of water and juice.

Pema tries to convince me to continue on to Upper Camp One, but I don't think he understands the severity of my condition. I feel bad for him, as he was carrying at least thirty kilos on his back (he had supplies for Upper Camp One as well as two oxygen tanks that were going to Camp Four). He continues to try to talk me into continuing, but I know that if I continue in this state, they will have to call a helicopter for a rescue, or worse, a body recovery.

I tell Pema that I'm pretty sure it's heatstroke, and I have no choice but to go back to basecamp, but he's reluctant. His unwillingness to support me in my decision stresses me considerably. I begin to feel quite ill again, then almost lose consciousness. As everything begins to black out, I put my head in my hands, take a couple really deep breaths, and prepare to wake up in a helicopter, or not at all. I manage to stay with it, though, and concentrate on breathing deeply.

I'm not happy with Pema, and don't want to be near him. He's now causing me grief instead of helping me.

"You go to Upper Camp One," I tell him. "I'll go back to basecamp on my own then, just leave me alone."

I slowly get back on my feet, even though I'm still wobbly. I know I need to head down, and just take deep breaths as I place one foot after another. Pema follows behind me, and I hope he just leaves me alone. Other sherpas coming up the trail would have radios, so I'll be able to make a call on the radio if I need to. I do realize that Pema can't leave me on my own as I am his client.

Going back down is slow. I need to pause often to stop feeling dizzy and like I'm about to black out. Every time I stop, it takes me about five minutes to recover. I reach the crampon point, take off my crampons, and hang them off my backpack. I don't even bother to take off my harness as I don't have the strength.

From there, I make my way slowly but surely to basecamp. I go right to the common tent and fire up the heater, as I'm feeling chilled. I'm so weak that I don't bother unpacking my bag or taking off my harness. I just sit next to the heater and feel like sleeping.

Our cook comes into the tent and asks me if I would like anything to eat. "Just some soup," I tell him. I'm not very hungry; I'm still feeling somewhat ill and a little shocky. After my soup, I sit there for about forty-five minutes, finally mustering up the strength to take myself and my gear to my tent for the night.

Saturday, September 16, 2017

I had a disrupted sleep last night, as I constantly woke up feeling dizzy and needing more oxygen. My breathing feels laboured; I really need to work to get my lungs to fill.

Then I remember that Samantha McMahon had told me that the world's leading doctor in high-altitude medicine was in the Chinese camp next to the helipad. Shortly after some breakfast, I tell the cook and Pema that I'm going to see this doctor. Pema decides to follow me, which doesn't matter to me either way; he can come if he wants.

Once at the Chinese camp, I ask to see the doctor. They are very welcoming and ask if I'm hungry or would like some tea. I thank them and say that I don't want anything. As I'm waiting, I notice what a nice setup they have here.

The doctor comes out to see me after tending to someone else. He checks my vital signs, and says that although he doesn't have an ECG monitor here, my heart sounds strong and regular. My blood pressure was a little high, and I tell him that whenever a doctor checks this, it usually shoots up and then settles down once I relax. He asks me what was happening and how I was sleeping.

After hearing of my symptoms, he says that I seem to be in decent shape and at night I'm suffering from high-altitude sleep apnea, causing me to wake up whenever I stop breathing. I should begin taking Diamox to help with this, half a tablet in the morning and another half tablet at night.

He thinks I got heatstroke on the glacier, and it's a good thing I came back to basecamp. The heatstroke seems to have triggered some altitude illness, though, and he advises me to rest in basecamp for a couple days before climbing again. He also asks that I come back to see him before I climb again.

I go back to basecamp and relax, enjoying the nice weather. I spend the day catching up on my journal and eating. I also talk with Ina and think that she's probably one of the strongest climbers on the team. When at home in Switzerland, she runs ultramarathons in the mountains. This explains why she is so strong and well adapted at altitude! Maybe I should start running—it would no doubt help me with the altitude.

We have dinner, sit around a spell talking, then I go to my tent for the night.

CHAPTER 12:
MY HEART-WRENCHING DECISION

Sunday, September 17, 2017

What a horrible night here in basecamp. I crawl into my tent and then into my sleeping bag. I pull my head into the bag and take deep breaths to help warm it up inside.

I fall asleep just before 8:00 pm and at about 12:30 am, I wake suddenly and sit straight up, gasping for air, panicking. I just came out of a dream where I was in deep water, drowning, and couldn't get to the surface to breathe. I'm so relieved when I realize it was only a bad dream.

I then realized I had woken up gasping because of the high-altitude sleep apnea that the Chinese doctor had told me about, but it hadn't been this bad before. As I lie there, I feel a chill throughout my mid back, as well as an increasing pain radiating from my left shoulder down through my arm and back up through my neck to my jaw.

At this point the pain is mild, and I don't pay it much attention, thinking it's caused by the chill in my back and the muscles cramping. I start to feel a squeezing pressure in my chest and around my rib cage. I initially brush it off as a mix of the cold and a bad sleep position; maybe I pulled a muscle in my back when I lurched up.

But the pressure and pain increase, and I'm feeling some anxiety as well as nausea now. I'm becoming very concerned, thinking I might be having a heart attack. I think about going to alert someone, exerting myself may not be a good idea, especially when it's so dark and slippery outside. Also, I don't want to wake anyone needlessly.

After a couple of minutes, the nausea and chest pressure ease up somewhat. I put the pulse/oxygen monitor on my finger; my pulse is 120 bpm and my oxygen level is at 68%.

I decide to take a couple aspirin and keep an eye on my pulse. I gradually feel better as my pulse lowers and my oxygen levels rise. But it sinks in that it would not be wise for me to continue the climb now. For the rest of the night, I am in denial, wrestling with the choice between trying to climb the mountain or going to a hospital to get my heart checked out. That would have to be done back in Kathmandu, as there isn't anywhere near here with the right equipment for this.

As the morning dawns, I succumb to the heartbreaking decision I must make. If I stay and climb with a heart condition, I will most likely not survive. Not only that, but this selfish action would also place others who may try to rescue me at risk.

Continuing would be foolish and very selfish. I must always remember that safety is paramount, and not let summit fever take hold of me. The mountain will always be here, and maybe I can come back one day.

I start to think about my loss now. The time and money invested. All my plans, hopes, and emotions. All of my hard work and shattered dreams. The feeling that I'm letting the youth and children down as well as Elton, my mom and dad. If I am able, I will need to come back and finish this.

There's also the expedition itself, my family here in basecamp, and my sherpa, Pema! I cannot find words to explain how I feel. It's like I'm in shock. I'm afraid I have hurt my heart not only physically, but spiritually and emotionally too. This is now out of my hands: I have no choice and will ask the sherpas to call for a helicopter.

As the Himalayan sun dawns over the snow-capped peaks, the light chases away the loneliness of the night. Yet it brings me limited comfort, because I know I must leave. I come out of my tent to see the spectacular hues of Mt Manaslu taunting me with its glorious alpenglow.

I enter the common tent, fire up the heater, then make myself a hot chocolate. The beverage is more warming to my soul than my bones. My Indian brother, Jitesh Modi, joins me in the tent. He says good morning and asks me how I am.

I tell Jite about my night and my decision to leave. He tells me of when he was on Mt Everest with a kidney infection, which worsened to the point of turning him around just short of the summit. Any physical illness or condition compounds at high altitude. Up here, sometimes the simplest of ailments can weaken you to the point of exhaustion and death.

Jite tells me I'm making the right and wise decision. Hearing this from him makes all the difference, relieving a great burden from my tortured soul. Soham Saigonkar, my other Indian brother, feels the same way as Jite, and also supports me in my choice.

Next to come into the tent is my newly adopted dear sister, Ina Condrea, the languages teacher, ultramarathoner, and strong climber from Romania. In the short time, that I have known Ina, I feel that we are kindred spirits, and my bond to her is like that of blood siblings.

When Ina learns what's going on, she saddens and tries and talk me out of leaving. I don't believe she realizes the seriousness of my situation at first. After a minute or two, she eases off and then becomes my soul's greatest source of strength, in a moment when I'm feeling so very alone. I am forever grateful to you, dear Ina.

Mathew Eakin from Australia also concedes that leaving is my only choice. It's comforting to hear this from him, an experienced and strong climber. I finally let Pema Sherpa know what's happening and ask him to call for a helicopter. I wish I could have stayed, as I could not have asked for better ascension mates on this expedition on Mt Manaslu.

The helicopter is on its way with Lakpa Thendu Sherpa onboard who will escort me to the hospital. I go back to my tent and pack up my things.

As I'm waiting in the common tent, Mingma Tenzi Sherpa brings me an oxygen tank and mask. He sets it up for me and I put it on, and the oxygen makes me feel better as I wait. After an hour, we get news that the helicopter has left Kathmandu. It's only a 45-minute flight and they should be here soon, so the sherpas gather my bags.

We walk slowly over to the helipad, which is beside the Chinese camp that overlooks the valley and Samagaun below. I'm only carrying my oxygen. Ina is directly behind me, along with Jite and Soham. As we are walking to the

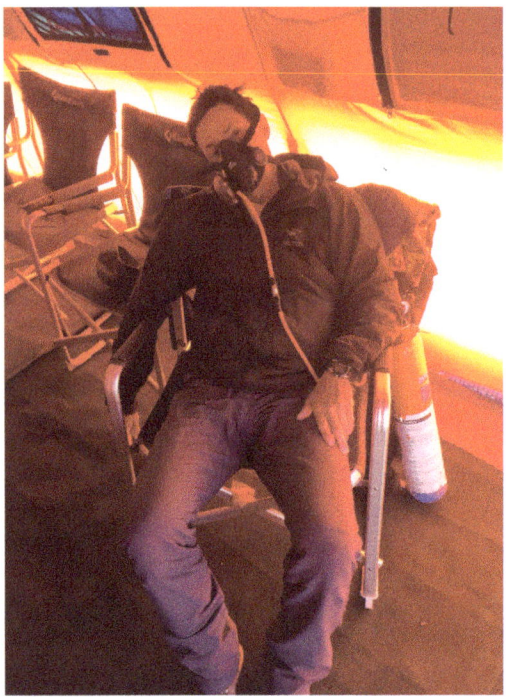

helipad, I feel embarrassed at how many people are watching my slow walk of shame.

Once there, I sit down with Ina at my side. We enjoy the morning sun and light breeze on our faces for about ten minutes. I then walk over to Pema Sherpa, say my goodbyes, and apologize for any inconvenience that I may have caused him. I see his eyes well up with tears, which I did not expect at all.

I have to leave these sad thoughts and begin to take a more positive note. What a beautiful morning: not a

Waiting for the helicopter with oxygen on.

cloud in the sky, and Manaslu's icy slopes glowing in all their glory as it seems they are beckoning me to come back.

We can now hear the helicopter approaching Samagaun. A minute later, we can barely make it out as it lands in the village to drop off supplies, its rotorblades flickering in the sunlight. Lakpa Thendu Sherpa will hop out of the helicopter and wait for me there, as the helicopter needs the room to bring more supplies up to basecamp.

After a few minutes, the helicopter takes off to begin its climb up to us. It makes a couple of wide sweeping circles as it fights to gain altitude in the thin air. Then it circles basecamp before landing on the pad next to us. I can easily see that the helicopter is an AS350 B3e, made by Eurocopter which is now Airbus Industries. (I worked on many of these when I was employed at Canadian Helicopters.)

Once the helicopter settles, the rotors still whirring overhead, the sherpas open the door and unload the supplies for basecamp, then place my bags in the back. I say goodbye to everyone there, and giving Ina a hug.

I get into the co-pilot seat with my oxygen and strap myself in, then Mingma Tenzi closes and latches the door from outside. He looks at me, gives me a thumbs up then gets clear of the helicopter.

CHAPTER 13:
FREE BIRD

Sunday, September 17, 2017

The pilot spools up the engine. He lifts off and banks left to get out over the valley. Once we have some altitude, he begins a steep descending spiral to the right to lose altitude. It feels almost as though we are free falling sideways. I can hear the rotor blades loudly chopping through the thin air, and my water bottle on the dash in front of my knees begins to float in the air a bit. It reminds me of going on test flights at home, doing autorotations and feeling like your stomach was coming up your throat.

(An autorotation is what you do to land safely in a helicopter when your engine fails. You essentially freefall while travelling forward with the wind making your rotor blades spin. When close enough to the ground, you pull up on the collective control, which increases the angle or pitch of the blades, to set your helicopter down as softly as possible. When practising these, you leave the engine on with reduced power and you don't actually land. At about fifty feet of altitude, you increase power to fly again.)

Once we land at the helipad in Samagaun, some locals refuel the helicopter with jet fuel from five-gallon jugs. Lakpa hops into the seat behind the pilot.

Everyone gets clear of the helicopter, and the pilot again spools up the engine, pulling the collective pitch control up. I see the rotor disk curl up so slightly as the blades take the weight of the helicopter. As we rise off the pad, he pushes the cyclic stick forward. The helicopter effortlessly obeys his every command. Our bird's nose drops and we begin to rise, sweeping over the village. Picking up speed as we climb out, the valley opens up to expose its vast expanse of deep lush green.

The pilot isn't a local, as he had blonde hair and blue eyes. I ask where he's from.

"Austria," he replies.

In my best Austrian accent, I ask him, "Ahh, yah, yah, like das Awnold Schwarzenegger, do you tellen zee volks to git to zee choppa?" He lets out a big belly laugh, and I laugh with him. We hit it off. I tell him of my years working at Canadian Helicopters, and all the fun we had testing the birds and doing autorotations out by Big Lake, just northwest of my home city of Edmonton.

Most of the way back to Kathmandu, In my head I have the song "Free Bird" by Lynyrd Skynyrd playing over and over again. The flight was scenic as we flew through the steep, narrow gorges topped off by snow-crested peaks, with the green of the jungle clinging precariously to their sides. Farmhouses with little terraced gardens are built in what seem to be inaccessible locations, with severe vertical drops to the valley below. I can only see a small trail connecting these houses, cut into the sides of the slopes.

We are now flying into the foothills that precede the Himalayas from Kathmandu. These foothills are as tall as our Rocky Mountains at home in Canada. As we fly, the lush green jungle covering these massive hills comes up to meet us as the terrain rises, then drops off the other side. The morning sun's rays evaporate the mountain dew into a wispy fog that rises into the heavens above, creating misty mountain tops and developing clouds. This flight is almost hypnotic—unless it's the altitude illness.

Like a bird on the breeze, we soar over one more hilltop. I can almost smell the colourful Himalayan flowers that race below us as they bask in the sunshine. Our rotor disk cuts a swath through the cloud being born of the dew. It opens up to us the Kathmandu Valley and its ancient and spiritual city of the gods.

Looking upon Kathmandu from above, I admire its ancient beauty, multicoloured houses, and lush green flora. In the near distance, I recognize Boudhanath Stupa as the prominent mandala encircled by its original neighbourhood. As we approach the airport, I check the outside air temperature gauge. It reads 33°C—what a difference from the mountains! We land at the bustling helipad of Tribhuvan Airport, where helicopters are coming and going like bees from their hive.

One of the ground crew opens my door; I thank the pilot and step out. Not only is it 33°C, but the humidity must be around 95% as it feels like I just walked into a steam room. I'm immediately soaked. Now I'm feeling woozy again from the heat, and one of the ground crew leads me to a luggage wagon that I can have a seat on.

Lakpa joins me and the ground guy brings me a bottle of water from a cooler they have on the tarmac. Right now, this heat is just too much. After a couple minutes, a van comes. We get in along with my bags, and it takes us around the airport to the terminal building.

On our way, we pass the hangar for the Nepali military. Standing guard on the tarmac are four soldiers with machine guns, watching over a couple helicopters and a four-engine turboprop airplane, all painted in camouflage.

The van takes us around the side of the terminal to a gate with a guardhouse, where a soldier checks our credentials. He opens the gate for us and we drive out, stopping at the public entrance to the terminal.

This is where we transfer to my chariot, (yes it's an ambulance). *Sure beats taking a taxi*, I think as a pretty nurse opens the side door, holding my arm as I step into the back. She asks me to lay on the stretcher. Then she hooks me up to the monitoring equipment. We race through the streets of Kathmandu with the siren blaring, *beedoo beedoo beedo!* That is, if you want to call it racing, as the traffic is so congested here.

I'm facing the back window and shoot some video as I chat with the nurse and tell her what happened. A couple guys on motorcycles are following us to get through traffic faster as we clear the way for them, this gives me a laugh.

We arrive at the Swacon International Hospital, where another nurse comes out with a wheelchair. I politely decline and tell her I'm okay to walk, so she then leads me to a bed in Emergency and asks me to lay down. She checks my pulse and blood pressure. I think this will be good for a laugh. She checks my pressure—wait for it, wait . . . then her eyes open as wide as saucers!

She leaves and comes back with the doctor, who checks my blood pressure. His eyes open right up in surprise, too, as my reading is 160 over 92. I chuckle and tell him to give it a few minutes, as I become tense when my blood pressure is checked and it usually shoots up.

He relaxes and hooks me up to the ECG and checks my heart. When he checks my blood pressure again, it's come down to 124 over 70. The doctor tells me I have White Coat Syndrome. Interesting—I didn't know it had a name!

He tells me my heart is strong and healthy, the chest pressure and shoulder pain I felt was more than likely from anxiety or cramping from a chill. Thinking back, I was chilled that night. I have a little altitude illness and will need to stay a night or two at the hospital.

They run me through a whole battery of tests and even check my eyes thoroughly. I then get to relax in my own room on the fourth floor. My room has a nice big television and its own washroom with a western style toilet and a shower.

My nurse comes in, takes my pulse, and gives me a couple pills to take, along with a dinner menu. She is gracious and polite; her name is Sukriti, and she is another Nepali beauty. Lakpa then comes up to my room and we talk for about fifteen minutes. Before leaving, he says he will come by tomorrow as I may be released then. I tell him I hope so—not that I don't like it here, but it would be nice to get outside as I feel caged up.

After Lakpa leaves, I have a shower and shave, which feels amazing. The only windows in my room are in the washroom. They're hazed for privacy, but if I open them I can see outside. Directly outside the hospital is the Dhobi Khola River. It's not very big, only about twenty feet wide and has brick walls on either side that are about ten feet tall. On the other side of the river is a typical inner-city neighbourhood, with brightly coloured multi-level buildings. It's dusty, like the rest of the city, and there are chickens clucking about.

After my shower, I have a nice dinner and then settle into my bed to watch some TV. I find a movie channel that is playing *The Mask of Zorro* with Anthony Hopkins, which I've never seen. Imagine that, I came all the way to

Nepal to see *The Mask of Zorro*! After the movie I get comfortable under the sheets and go to sleep.

Monday, September 18, 2017

I wake up at 6:00 am this morning in my cozy little hospital room. My blood pressure is 122 over 68 with a pulse of 59 bpm when Sukriti takes the readings asks what happened to me and I tell her the story. The doctor then comes and checks on me and says I'll be discharged today, sometime after 10:00 am, which makes me happy.

After the doctor leaves, I eat breakfast and have a quick shower, then get ready to leave. As I'm getting ready, I see myself in the mirror and realize how much body mass I lost on the mountain. When I get back home, I will have to get back to weight training, as well as training with my ice axes for mixed climbing. I'll also need to do specific high-altitude training if I am to come back and finish what I've started.

Just after 10:00 am, Sukriti gets my discharge papers ready, then Lakpa and his brother Pemba come to pick me up. Before I leave, the doctor says he wants me to come back in a day or two for a follow-up exam.

Lakpa, Pemba, and I get a ride back to the hotel in the hospital's van (I think the ambulance was faster). Back at the hotel, Nabin and his family are happy to see me, and I feel the same. I have an even bigger room this time on the fourth floor. I sit for a while and chat with Samir. I'm tired and still recovering, so I have a nap on the big comfortable bed in my room. It's almost as comfortable as my Donald Trump mattress at home. You heard me right: I actually have a Donald Trump mattress! Believe it or not Trump had a line of home furnishings named after him.

After my nap, Samir says he needs to go to a local bakery to get some items for the hotels café and asks me if I'd like to come with him. I say sure, and we leave.

The bakery isn't far away; it takes about eight minutes to walk there through all the hustle and bustle of Kathmandu. We turn off the street into a narrow walkway between two ancient brick buildings where I begin to smell the bakery. Children are playing everywhere in the narrow walkways. We make another turn and go into the back door of the bakery. I could have died when we entered, as all I can see are cakes, pies, and pastries filling the cooling racks. The aroma in there makes my mouth water. What a busy little bakery!

Samir gets some fresh bread, buns, and a few pastries, then we're on our way back to the hotel. I go to my room and watch a little television, then get ready for dinner and go up to the café on the roof. I have dinner then chat with Samir and Aashish as I have some coffee.

Well, off to my room and then to bed for the night.

Wednesday, September 20, 2017

It's so nice to awaken to the sound of roosters in the morning again, especially in a city of this size (officially there are about 2.5 million people here, but it feels like it's considerably more). My sleep was good; I shower and go up the marble stairs to the roof for breakfast amongst the small birds and sunshine.

A buddy of mine from San Diego, Ash Gambhir, calls me on the phone, and we talk for about six minutes. It's nice to hear from him. Then I see on Facebook that Reinhold Messner, a famous climber from Italy, is in Kathmandu. *Wouldn't it be cool to bump into him?* I think. Reinhold was the first person to climb Mt Everest without supplemental oxygen, and he holds a few other records, too. After some coffee, I go out to shop for friends at home.

As well as getting some souvenirs for others, I bought myself a nice paint- ing of Mt Ama Dablam. In the foreground, it features a sherpa and two mountaineers crossing a suspension bridge over a Himalayan river. The mist rises out of the valley behind them to expose this most beautiful peak in all its splendour. I also got myself a hand-carved wall enclosure containing three prayer wheels you can spin for blessings and good luck.

I'm back at the hotel now waiting for Lakpa. We're going to take the gear back to the trekking store, and I should get some money back as well.

We have milk tea at the trekking store and visit for a few minutes. Hanging on the wall is an old ice axe from about 1920-1940; I inquire about this axe and it's price and then I bought it to hang on my wall at home. After this, Pemba and I will go back to the hospital so I can do my follow-up with the doctor. I also need the doctor to sign my insurance papers so the cost of changing my flights is covered.

On the way to the hospital, I ask the hotel's driver if he can stop so I can pick up some flowers to thank Sukriti for taking care of me. He stops in front of a small flower shop that appears to be owned by a husband-and-wife team. It's dusty inside with many plastic plants. As Pemba asks about fresh flowers, I think that it looks like they don't do much business here they might not have any nice flowers.

When the woman comes out of the back of the shop with some absolutely beautiful red, pink, and white flowers, with some red baby roses mixed in, I know we're in the right place. They don't have any appropriate cards, and what they do have is covered in dust, so I just pick out a nicely decorated golden vase and we are on our way.

At the hospital, the doctor checks me over and says I'm recovering well, then he signs my paperwork. I ask if Sukriti is upstairs in her ward, and he says she isn't there. Instead, I give the flowers to one of the other nurses there in emergency and ask that they give them to her for me. They then want to take pictures with all the nurses, the flowers, and me.

As Pemba and I are about to leave, and I say to please make sure that Sukriti gets her flowers, the doctor says, "Sukriti! She is upstairs. I will call her." I guess he didn't understand me when I first asked for her.

Sukriti comes down from upstairs and around the corner with a puzzled look on her face. Once she sees me and the flowers, a big smile replaces her confused expression. We take a couple pictures and I thank her for taking such good care of me. We become Facebook friends, I say my goodbyes, then Pemba and I go back to the hotel.

At the hotel, I go upstairs to the café to have a snack. It's another beautiful day, so I decide to take a walk around the Thamel District.

Too funny: as I make my way over to what used to be called "freak street" in the sixties and seventies, where all the hippies would buy and do their drugs, I notice my purple sunglasses have a curious power. When I'm wearing them, plenty of guys come up to me asking if I want any hash or weed. But when I take off my purple shades—nothing. I put them on again, and the drug dealers are back! So if you are in Kathmandu and want some weed or hashish, just wear purple sunglasses on freak street. The weed will find you! (By the way, freak street has transformed from a drug hub to a tourist hub now.)

I walk the other way, back past my hotel and down by the Garden of Dreams. As I sit on the corner watching traffic and people, minding my own business, an older man comes up to me. I think he's a yogi or a holy man. He chants and put a *tika* on my forehead. (The *tika* is the red spot on the forehead of Nepali people for good blessings, health, and long life.)

"*Dhan'yavad*," I say to him, and give him five rupees. I get up and begin to walk back, this time along the wall that encloses the beautiful Garden of Dreams. The road is very busy and dusty from the traffic.

Just ahead of me, I see an Indian woman sitting on a blanket next to this busy, dusty road. As I get closer, I see that she is tiny and has a baby with her, swaddled in rags. The baby's face looks thin. How sad! It looks like she is a beggar.

When I get closer, she holds her hand out towards me, but does not utter a word. Up close, I can see that she is frail and emaciated. Of course, this tugs at my heart strings. I reach into my pocket and pull out 1,000 rupees (about $10 USD) and place it in her hand. As I do this, she wraps her warm tiny fingers around my hand, then with tears in her eyes she says, "*Dhan'yavad mero sathi*," meaning, "Thank you, my friend."

After this, I head back and eat dinner at the Gaia restaurant by my hotel. I think of the poor woman with her baby, and feel so grateful for all my blessings. I have learned that many people come here from India to panhandle for survival. In many cases the man in the family dies, quite often in a work related accident and leaving the children's mother to beg for money or food.

September 24, 2017

It's my last night in Kathmandu, as I'm headed back home tomorrow. The past few days have been uneventful. I have just been relaxing and people-watching in the Thamel District, and also talking with the local people as they are so friendly, warm, and engaging. Tonight Lakpa, his brother Pemba and I will go out for dinner to a Nepali cultural centre.

I'm back in my room after a great night. Lakpa, Pemba, and I went to the Nepali cultural dinner theatre called Bhojan Bhumi. It's right around the corner and is very busy. Here we have great food with some Nepali wine, which is basically moonshine. It's very strong, so you only sip it. During dinner, musicians and a singer play for us as dancers perform traditional folk dances, such as the Sherpa Dance and my favourite, the Yak Dance. For the Yak Dance, two performers in a furry black yak costumes dance around the tables and collect tips in the yak's mouth.

September 25, 2017

I had a good sleep last night with a belly full of yummy Nepali food and a glow on from the wine. My flight boards about 7:00 pm, so I will leave for the airport around 4:00 to 4:30 pm.

I have breakfast upstairs with a solemn feeling, as this is my last day here for some time. Young Samir got me a going-away memento, a little fuzzy red bird keychain for which I am thankful. The rest of the day I go for a couple walks just to have a last look around for memory sake. I say my goodbyes to everyone at the hotel, then we depart.

On our way to the airport, I feel like I am leaving my family. Everyone has been so good to me here. I say goodbye to Lakpa at the airport as he places a Tibetan prayer scarf around my neck.

I will miss all of you dearly until I return. *Dhan'yavad mero pyaro bhai's! Meaning, thank you my dear younger brothers*

Namaste!

PART 3:
FACING MT MANASLU

CHAPTER 14:
SUFFERING OF THE SAMA AND HAPPY RETURNS

Sunday, September 2, 2018

Waiting for my flight out of Edmonton, I can hardly believe I'm actually on my way back to Nepal and Mt Manaslu. When I got home last year, I felt that my soul had been touched somehow by the people of Nepal and the children of Samagaun. I began to do some research on them. From a video on YouTube called "Documentary on Samagaun Village at the foot of Manaslu," I learned their living conditions are very difficult, as I had suspected. Four out of ten children there die before the age of five, due to their remoteness, harsh conditions and related illnesses.

I would like very much to do something to help the children in Samagaun. Maybe there's a reason I'm returning to this village, and maybe this reason is much bigger than climbing Mt Manaslu. I remind myself how important it is that I complete my mission to help children, in memory of my friend Elton and my folks.

Since Samagaun is right on the Nepali/Tibetan border, the people there are all but forgotten by either government. There isn't much medical help as there are no doctors, maybe a couple of locals trained in some nursing at best. Samagaun is remote and difficult to get to: no roads that lead there, only trekking trails that take seven to eight days, like the one I travelled last year. It only takes forty-five minutes by helicopter, but this is costly and highly weather-dependent. Outside of climbing season, which is only a couple of months, helicopter traffic is virtually non-existent as this leaves ten months without helicopter flights.

In the winter they get a lot of snow and it gets quite cold. The homes have no heat. They are damp, cold, and full of smoke from the wood stoves used for cooking, which leads to many respiratory illness and often death from pneumonia. Malnutrition and sanitary issues also cause disease. Many of the people live as their ancestors did hundreds of years ago.

Yet the people in Samagaun are open, happy, and loving. Most here would open their hearts and homes, sharing their meagre amount of food with you if you were hungry. I feel this has much to do with their Buddhist beliefs, which heavily depend on being compassionate to other life. In large part, this is how they get closer to their heaven, or *nirvana*, when they die. After enough cycles of reincarnation, they achieve perfection and go to their heaven. Being kind to others enables them quicker transport to their souls' residence with their gods in heaven.

Even though they suffer to scrape out an existence at the foot of Mt Manaslu and endure the tragic loss of many children, the people of Samagaun still find it in their hearts to be unconditionally kind to others like me. For this, I am forever grateful, and have learned more from these people in a very short time than I have learned in all my life. Because of this, I feel my failure to climb Manaslu last year has become a success, as I will return to Samagaun in a week. I've decided this time I will bring a gear bag crammed full of new stuffed toys. I would like to do more to help the children of Samagaun, but this is all I can afford for now. My hope is that these stuffed animals will bring a little comfort to the sick children there.

Once back in the village, I will talk with Lama Guru at the monastery and see if he will keep the toys there, to be given to any children he feels would get comfort from them. Before I purchased the toys, I visited a monastery in Edmonton and asked the opinion of the head monk there. He said this was a great thing to do, thanking me for my compassion and inviting me back anytime I wanted to visit.

Maybe in the future I can come up with enough money to supply a couple helicopter flights yearly to fly a doctor to the village, but I will need to find a doctor in Kathmandu who is willing to donate his or her services for a day or two. My original reason for climbing the Himalayas was to raise funding and awareness to help children at home, but I wish to also do what I can to help the children in Samagaun, too! This is my dream, in remembrance of my dear friend Elton as well as my mom and dad.

When I returned home last September, on Facebook I became good friends with Sarmila Giri. She's the sister of Nabin, the manager of the Sacred Valley Home Hotel where I stayed in Kathmandu. I now consider Sarmila my best friend, and after ten months I will be meeting her for the first time in person. I am very excited just for this. She has a master's degree in economics and lectures at two colleges in Kathmandu. I'm bringing a few presents to her for her birthday which is on September 20, although unfortunately I will be in basecamp on September 20.

I also struck up a friendship with Samir, the young man I went to the bakery with last time I was in Kathmandu. He invited me to spend Dashain, an important Hindu festival and family holiday, with him and his family. His home village is called Panchkal, a small farming community, a two-hour bus ride from Kathmandu.

Samir has since been accepted at business college in Sydney, Australia, and unfortunately will not be home for Dashain. Sarmila will be spending the holiday in Panchkal, though (Samir is her nephew and this is where she grew up), so she graciously invited me to travel with her. I am grateful to Samir's family for this invitation; I will stay at their house and share in their culture. I am very excited for this!

In March, I began training specifically for success on Manaslu. Fifty-four years old and now I am into running! It took long enough, as I never liked or needed to run or jog, even when I was bodybuilding. Now I was doing high-intensity interval training to increase my VO_2 max (for "volume of oxygen", also known as maximal oxygen consumption). I also kept up with weight training, reducing my weights for more reps and endurance, rather than brute power. I partially tore my upper biceps tendon, but it healed up well with some physiotherapy.

Eight weeks before leaving for Nepal, I rented training equipment from Hypoxico in New York City to pre-acclimatize for high-altitude conditions. I slept under a high altitude head tent set up on my bed, reducing my oxygen levels gradually until I was sleeping at the equivalent of 20,000 feet. For two hours daily, I intermittently breathed oxygen equivalent to 21,000 feet, and every two days I rode the stationary bike with the a Hypoxico mask on, working up to about 16,000 feet altitude while riding.

After eight weeks of this training, I'll be going to Nepal pre-acclimated. The only tricky part is that once I stop hypoxic training, my physiology begins to return to normal and the high red blood cell count I worked so hard to achieve starts to come back down. Like the saying goes, *use or lose it!* The sooner I get back to the mountain and its altitude, the better off I will be.

Because of this, I will not be trekking into Samagaun this time, but flying in from Kathmandu to Samagaun by helicopter. Once there, I'll climb up to basecamp as soon as possible.

I sure hope my hypoxic training works. I'm a little concerned, having just read about a British climber who got altitude illness in Manaslu's Camp Four last year. On his way down to recover, he died between Camps Three and Two. This was just a week after I had left the mountain. Godspeed to him and prayers for his family.

* * *

I'm back at the airport now, about to begin my long journey to back to Nepal. Here's a shout-out to the front line ladies here at the airport for flying the bag of stuffed animals for humanitarian reasons, without charge. They also didn't charge me for my overweight bags (although though they were not far over the limit).

It's funny, every time I think of what I am doing, I get a rush of excitement and giggle—but at the same time I feel some trepidation, considering what happened last year. Luckily, I was able to get window seats all the way from here to Kathmandu; watching the world pass by also passes the time.

Like last time, before leaving for the airport I stopped by my mom, dad and grandma's resting place at the Mt Pleasant Cemetery. I chatted with them, asking them to be with me in spirit and enjoy the journey through my eyes.

My flights are good. I love flying across the Pacific Ocean with Cathay Pacific Airlines: the food is out-of-this-world yummy and the Boeing 777 Dreamliner is very comfortable. I even have three seats to myself to stretch out. I also think it's so cool to be travelling at ground speed of almost 700 mph across the Pacific Ocean.

This time, Hong Kong is smog-free and sunny. It's just astounding how big the city is, with too many skyscrapers to count. When we leave Hong Kong, it's already dark due to another long layover. As I peer out my window down on the water, I see evenly spaced white lights, hundreds of them for as far as I could see. I ask the stewardess what these are, and she says they are fishing boats. *I sure wouldn't want to be a fish down there*, I think.

Yay, here I am! Namaste from Kathmandu! I arrive just after 11:00 pm to the familiar humidity and heat. It feels so wonderful and welcoming to again hear the night song of the crickets as I walk to the terminal. It instantly brings a smile to my face, even in my zombie state. I feel like I've arrived home.

September 5, 2018

Good morning! I had a wonderful sleep, and how I have missed waking up to the roosters and songbirds. Oh yeah, and also the crows, *caw caw caw!*

I eat a tasty breakfast with my *bhai* (younger brother), Lakpa Thendu Sherpa. He tells me that I will be flying to Samagaun on September 7, rather than the 10th. This is great news for my pre-acclimatization, but these next two days will be hectic. Today Lakpa and I will go to the trekking store to get the gear I need, and tomorrow I will pack for the trip. Tomorrow I would also like to meet my friend Sarmila after she is finished at the college, if that's okay with her.

* * *

We've just come back from the trekking store (Chhiringma Trekking Shop), a nice visit with hot milk tea. They didn't have boots in my size or the snow goggles, so I will come back tomorrow to pick them up.

After dropping the other gear off in my room, I go out for a walk. I'm happy to see my favourite dog lying in her usual spot, just outside the coldstore down the street. As I'm walking towards her, she recognizes me, and gets up to meet me. I can tell by her wagging tail that she's happy to see me again! I get down on one knee and spend a good ten minutes petting her and scratching behind her ears. Looks like she's lost a bit of weight—good, because last year she was too chunky for a dog.

I see the Holy Himal hotel across the road from the Fusion Café has finished building their new restaurant. The addition is brickwork, with a rooftop deck, statues, and nice big ferns outside. It looks great!

Next to my hotel is a shop, or maybe just a gallery, of Nepali/Hindu brass artwork. Displayed in the window is an incredibly intricate brass statue of Shiva that stands five feet tall. I try to comprehend all the work that went into creating this but it is so large and intricate that it baffles me.

I walk around a little more, then decide to head back to my room for a rest. With such a big time change, the jet lag is a butt-kicker. After my nap, I go down to the lobby and chat with Nabin and the family. It feels a little empty without Samir here, but he is doing very well at college in Sydney as I stay in touch with him online.

My furry, well-cared-for friend on the step near the cold store.

Incredible craftsmanship in this brass statue of Shiva.

If you remember, Purnima Shrestha, a photojournalist from here in Kathmandu, was with the Women's International Team on Manaslu last year.

Since then, she and I have adopted each other as brother and sister. She was successful on Manaslu, her first mountain ever, and this past spring she also summited Mt Everest. I am so proud of her; she is a rock star!

Now that I'm in Kathmandu, she comes to see me and brings me an Everest-related T-shirt that says, "Top of the World, Nepal." Purnima and I talk for a few minutes at a table on the hotel's front veranda, then we go down the street to the Gaia Restaurant for a snack. I learn that her reason for wanting to climb Everest was because someone said that she wasn't capable. She then turned the climb up Everest into a great cause by bringing awareness to cervical cancer in Nepal and educating youth to get vaccinated against HPV.

She is such a beautiful soul, as well a very talented photographer and pretty too. The man she chooses to marry will be blessed by God. I only hope whomever he is, he loves her unconditionally and is true to her sweet heart.

After coming back from meeting with Purnima, I hang out around the hotel, chatting with Aashish, Nabin's son, before retiring to my room. I get ready for bed, still fairly tired from my flights.

September 6, 2018

I'm still adjusting to the time change, as I had a disrupted sleep last night. I woke up at 1:00 am and wasn't able to get back to sleep until almost 4:00 am.

Just had my shower and I'm getting ready to go upstairs for breakfast. Today I'm also going to start taking Diamox and ginkgo biloba to help with the altitude. I'm not taking any chances this year, as last year was pretty expensive.

I will never tire of the beautiful mornings here: 8:30 am and it's sunny with a nice breeze, the birds singing, and it's 24°C. Nepal has 9% of the world's bird population with 866 different species recorded. If you are a bird lover, this is the place to be. My mom was a bird lover, so hearing and seeing all the little birds here draws me into feeling close to her spiritually. I know that she would have loved it here.

I have my breakfast and Lakpa joins me just after I finish eating. We go back to the trekking store to get the items I still needed. He tells me that his brother Pasdawa will be my personal sherpa for Manaslu this year.

* * *

Back from the trekking store, I pack all my gear. Around 4:00 pm, Sarmila comes to the hotel and we finally meet each other. Both of us are a little shy to begin, but we expected this.

Sarmila and I joined Nabin and his wife, Laxmi, for a snack upstairs at the café. I am intrigued by Sarmila and find her very attractive. I look forward to spending more time with her after Manaslu.

Sarmila, her niece Se Le Na, and I leave the hotel and walk to Sarmila's apartment, which isn't far from Swayambhunath (the Monkey Temple). Se Le Na is Nabins daughter and is a very personable and polite young lady. It takes us about fifteen minutes to walk there in the dark, as we pass through some older areas of Kathmandu and out of the tourist zone. Many of the locals look at me as if to say, "Are you lost?" until they realize I am with the ladies. Dogs and children are everywhere, as well as little stores with their goods hanging outside, bakeries with their yummy treats on display, many butcher shops, and people selling fresh food from vending carts in the streets. As we pass by these food carts the delicious aroma makes my mouth water. I love the atmosphere here. To me this whole city feels like a very large ongoing festival. So many colours, lights and friendly happy people.

We then cross the Bishnumati River, which looks as though its water volume must have been much greater in the past. But this is common around the world, as many rivers and streams are drying up.

A few minutes later, we arrive at Sarmila's place. It's very cozy, tucked away from all the traffic and bustle so that it's nice and peaceful. I've brought Sarmila's birthday presents and a couple other things she asked me to bring, like honey and some lipstick. I brought some wrapping paper and bows from home so I could wrap her presents, which I did at the hotel last night.

I pull the gifts out of my backpack and say that she can wait for her birthday or open them now. I should know not to give a girl nicely wrapped presents and expect her to not open them, correct? It was nice to see the expression pleasure on her face. I got her a nice bottle of Givenchy perfume, a big stuffed bear, chocolate, and a couple other small things.

We chat and eat some sliced apple and *padmas* (roasted soybeans—they are very crunchy and addictive), then we go back to the hotel. Once there, Nabin joins us and we go to the Gaia for a late dinner. Nabin treats us and it's very tasty. We make it just in time because the kitchen closes at 9:30 pm, and we arrive there at about 9:20 pm.

Rather than go all the way back to her apartment, Sarmila has brought her work stuff with her to the hotel. She'll stay in a room with Se Le Na and leave for the college from here in the morning. I will not see her in the morning, as she will be leaving at about 5 am. So before going to my room, I say my goodbyes until I return from the mountain.

Friday, September 7, 2018

I had a great sleep last night and just have finished breakfast on the roof. Lakpa joins me and tells me that we will find out sometime after 10:00 am when I need to be at the airport. He will also take me to exchange some U.S. dollars for Nepali rupees.

Purnima texts me to say she would like to get a couple pictures of me before I leave; I tell her okay, but that I'll probably be leaving a little after 10:00 am. I now have the rupees in my pocket and am waiting in anticipation.

We just got word! We will be leaving for the airport at about 12:30 pm. No word from Purnima by the time we head out, though—too late now as we must leave. Lakpa is escorting me to the airport and waiting there with me. On the way, we stop at a market and Lakpa runs in for a bag of fresh tomatoes that I will bring to Samagaun for the camp cook.

At the airport we go through the main terminal to the back where the helicopter charter companies stage their customers and baggage. It's sweltering hot in the airport, and we are told we might be waiting for some time as there are clouds building up in our flight path through the mountains. Helicopters in the Himalayas cannot fly in heavy clouds or at night because of the likelihood of flying into the side of a mountain—the valleys are very narrow, essentially they are deep gorges carved through the mountains. If the weather doesn't get better, I might be returning to the hotel for the night and trying again tomorrow.

Finally, four hours later I have parted company with Lakpa and am now in the shuttle van going to the helipad. As we get out of the van I see a familiar face—it's Arnold, Arnold "git to zee choppa", the Austrian pilot who rescued me from basecamp last year. As soon as he sees me, he laughs and waves; it turns out that he is our pilot.

When we're about to depart, two guys jump into the two-person jump seat up front, so I hop into the back seat behind the pilot. We chat for a bit and reconnect before he fires up the helicopter. The helicopter is full, with six of us on board besides the pilot.

In the helicopter about to leave for Samagaun, sitting beside the young Chinese guy with his GoPro.

Beside me is a young Chinese guy with his GoPro camera on a stick. He has no sense of safety, or just doesn't care. As we're flying, he keeps sticking his camera over the pilot's shoulder and in his face, I think to film the instrument panel. The pilot even knocks it out of his line of sight a couple times, and this idiot doesn't get the message. I've got to give the pilot credit for his patience, as I would have grabbed the camera and put it where the guy couldn't reach it until we landed.

Flying through the Himalayas is like nothing else: the views were absolutely stunning, with so many waterfalls and stone houses placed precariously on the lush green sides of steep slopes with sheer drop-offs. It's an interesting flight. Our pilot has his hands full as the clouds are building up en route, forcing him to do some cloud dodging and extra turns. He's trying to keep to light, wispy areas of the clouds so he can maintain full visibility of the ground.

We've been in the air forty-five minutes, and I can see the familiar blue roof on the Sama School where many children live and attend classes. (Because many of these kids are orphans, they stay at the boarding school.)

On our downwind leg past the helipad, we fly along the tree-covered slope along the back side of the village. Near the monastery, we bank steeply to my side to turn for our final approach, then land upwind and softly on the helipad.

The pilot looks back at me and we give each other the thumbs up. We all hop out of "zee choppa." A couple sherpas pull our bags out, then we clear the pad and film our bird leaving. She rises swiftly into the Himalayan sky and disappears into the distance of azure blue.

CHAPTER 15:
HEAR TELL OF THE HEADLESS DOG—RUFF, RUFF, RUFF!

Samagaun, Gorkha, Nepal. I had no idea how much I've missed this wonderful little village and the souls who reside here.

I'm now at the Norling Hotel. I have a comfortable room and there is a restaurant on the second floor with a sundeck and great views of Mt Manaslu. I get settled into my room and go for a walk.

As I get near the Tashi Delek Hotel, where we stayed last year, I can see they are constructing an addition with about eight to ten more rooms. The stone fence where they slaughtered the goat and the small building with the shower room and washroom is gone. With the addition, the hotel now forms a C shape around the courtyard.

Outside the front stone fence, two women and a young man are chiselling stone into the shapes required for the construction. Two children are playing nearby. I wave at them and beckon them over to me. I give them each a candy, as well as to their mother and the others hammering endlessly at the rock.

On the roof of the new addition, five men are mixing, forming, and setting the cement and stones. For shoring, they are only using hand-stripped logs. The new addition looks great, and the second-level rooms look ready to be occupied. Hopefully they've solved the rat and dog problem!

New addition to the Tashi Delek Hotel beside the tricky old wooden stairs that angle back.

I see the owner and his wife, and go say hello to them. He tells me the rooms on the ground and second floor are almost ready, and the men on the roof are adding a third floor.

The Norling Hotel with pack mules resting for the night.

I continue walking and find it so heart-warming to see all the children with their little rosy cheeks chasing each other around. At the same time, it's also heartbreaking to know of their struggle to survive here, especially through the cold, wet winters. I have just enough Werther's Original candies left in my pocket, and give them to these kids.

I finish my walk by taking some video and a couple pictures, then head back to the Norling Hotel for dinner and to relax. I enjoy a tasty plate of spaghetti with melted cheese, then enjoy a tea and catch up writing in my journal. The bag of stuffed animals should be here tomorrow. In the morning, I will go up to the monastery and ask Lama Guru about leaving the toys with him, so he can give them to the sick children as needed.

Saturday, September 8, 2018

I had a very good sleep last night and there was no dog barking, yay! I already love this hotel.

The only drawback is there is no Wi-Fi or cell service. I ask the guys here and they say there's a problem with the antenna or tower in the village of Arughat that sends and receives the signal here in Samagaun. Hopefully they get this fixed soon, as I would like to send out some pictures and a couple messages to some friends.

I go out onto the roof deck and look up at Mt Manaslu. Its twin peaks gloriously reflect heavenly white light from the sun against the deep azure backdrop of the Himalayan sky. I could stare at it all morning long.

I need to talk with Lama Guru so I trek up to the monastery. On my way up the trail, I meet a few pack mules, free from their burdening loads, grazing as they come down the trail. When I reach the monastery, it's closed, and there is no one nearby. I am disappointed, as I need to speak with the Lama before I leave for basecamp. I rest for a minute and check out the craftsmanship that went into building this monestary then turn around and go back down to the village.

As I'm walking near the Tashi Delek Hotel, I see Lama Guru! I say hi to him and tell him that I would like to talk. He gestures for me to follow him, and leads me into the courtyard of the hotel. He gets a local to be an interpreter, as he does not speak English and I don't speak Tibetan.

I tell him through the interpreter about the stuffed toys, and ask him if I may leave the toys with the monastery so they can be given to sick children to help comfort them. He says he will help me with this, asking that I let him know when the toys arrive.

Walking back to my hotel, one of the children I gave candy to sees me from where she and her brother are playing with some other kids. She stops and waves at me with the most beautiful smile you could imagine. Even with her hair messed up from playing, she is nothing short of adorable.

Back at the hotel, I'm sitting on the stone fence and basking in the warm sun. About fifty feet away, in the centre of the village, is that same stupa with the ornate paintings of Buddha inside. Some locals begin gathering by the stupa, and I see that some of them are distraught and crying.

Something has just happened, and it has a really bad aura about it. Two younger village men join me on the fence, and I ask them what is happening. They tell me that one of the helicopters that flies here just crashed, and they have lost communications with the it.

Oh, my God! My heart sinks as I hear his words. With the language differences and difficulties in translation, I thought at he meant a helicopter on the way here, but actually it was a helicopter that just left Samagaun. Now I really wish that I could get a message out to let people know that I'm safe and it wasn't our flight that crashed. I'm sure it won't take long for word to spread of this tragedy in Kathmandu and back at home.

The next couple hours go by slowly as we awaited news. So many scenarios run through my head, since I worked on helicopters for years. With the way our flight went yesterday and the clouds building so fast, my first thought is that the clouds built up around the helicopter leaving them with no escape. Maybe they had no choice but to set down on a slope, and ended up rolling down the hill into the valley, or they flew into the side of a mountain because of clouds blinding the pilot.

The fact that they lost communications suggests they crashed, because even if the helicopter landed in a deep gorge there would still be radio contact. The helicopter's Emergency Locator Transmitter is probably sending out a ping to satellites, which only means the worst—a crash. Oh man, what a way to ruin a beautiful day. I sure hope they are okay.

A couple hours later, we get an update on the missing helicopter. It was Altitude Air's AS350 B3e helicopter with the registration 9N-ALS. This was the helicopter beside us on the helipad back in Kathmandu. I then realized that I took video of the doomed helicopter on its last flight as it took to the air that morning. They were on their way to Kathmandu from here in Samagaun. After about eight minutes, they apparently flew into the side of a mountain; it sounds like the clouds enveloped them, as I had suspected.

Everyone died in the crash except for a female monk. We hear she is in critical and may not survive. Besides the pilot, there were six souls on board: the female monk, a Japanese trekker, and five Nepalis, two or three of whom were sherpas. This explains the distress here, as some of people who perished were locals. Godspeed to them, bless their souls and their families.

On top of the bad news of the crash, I have been told we are moving to the Tashi Delek Hotel. The first thing that crosses my mind is, *Wouldn't it be my luck for that bloody dog to be barking at the rats all night again?* It's too bad—I was quite enjoying the Norling Hotel.

I pack my things and take them across the village to the Tashi Delek. My room is right next to the room I had last year, funny, huh? In the dining room, I see the owner and his wife, and they seem happy to see me back here. I actually love it here—it was just the dog's barking that ruined it. For lunch, I have fresh cream of mushroom soup that is the best soup I have ever tasted in my life.

I've now met the other team members. Radu Albu is from Romania, where he runs his own expedition company. François Matter from France is a sports doctor, and has quite the story behind him. Not that long ago, he was climbing Mt Shishapangma in China, and an avalanche swept over him and his partner. His partner must have thought he was dead and left him there. François spent the night buried unconscious in the snow. The next day some one noticed part of him sticking out of the snow, then dug him out and rescued him. He's lucky to have survived, although he lost toes and two or three fingers to frostbite. Both of them seem very nice.

That's it this year for our camp—only three of us. In another camp, but still with Adventure 14 Peaks, is Taro Yamagata from Japan. I like Taro: he is very intelligent, has a lot of climbing experience, and he's also a very witty guy. Also under Adventure 14 Peaks is Leonardo Proverbio from Argentina. He plans on skiing down from the summit.

Interestingly enough, there is also Yasushi (Stan) Kuwahara from Japan. He was on last year's climbing permit but I didn't see him, so something must have prevented him from coming last year (I forgot to ask him the reason, though). Stan is a little older than me, and I like talking with him. He is also very intelligent and listens more than he talks, which says a lot about how wise he truly is. This year, he is on his own with Da Dendi Sherpa of Glacier Himalayan Treks and Expeditions.

As well, there is the very beautiful Vanessa Estol from Mexico City. Vanessa's guide is the famous Kami Rita Sherpa, who is currently the world-record holder for most ascents of Mt Everest—he has been on Everest's summit a mind-boggling twenty four times! It's not uncommon to find Everest sherpas here on Manaslu—many of them come here in the autumn.

Benjamin Treble, a friend of Mathew Eakin from last year and also from Sydney, Australia, is here with a buddy. I'm not sure what team they are climbing with. Great guys! I love the Aussies, and Benjamin brought me some pure zinc sunscreen that Matt had given to him for me. Thanks, mate!

Today, I also met my personal sherpa, Lakpa's brother, Pasdawa. He seems very nice and laid back, and I already feel comfortable about him being my sherpa. I sure hope he will not be in a hurry on the mountain. I guess I will soon find out.

Then a couple buddies from last year walk into the dining room: Mingma Tenzi Sherpa, big Mingma Temba Sherpa, and my personal sherpa from last year, Pema Sherpa. It's good to see them.

After lunch, I spend some more time out in the village taking pictures and handing more candy out to the children. This year, I notice a lot of construction and many new stone houses being built. Since the permit costs have increased for climbing Mt Cho Oyu, more and more people have been coming to climb Manaslu. It's getting very busy here in Samagaun.

I wish there were an extra charge in the permit cost directed specifically to Jigme Lama, who heads up the Sama Foundation to help people here. Considering that climbers pay an average of $11,000 USD, I don't think they would even notice an extra $100—but it would make such a difference to the people in Samagaun. The money could be used to fly doctors here, or supply simple potbelly stoves to heat people's homes. It could maybe even build a medical clinic for the whole area and train a couple locals as doctors or at least emergency medical technicians.

Sunday, September 9, 2018

Last night, I had a good time before going to bed. I was in the smoky kitchen with all the sherpas drinking a concoction called *chung*, another strong form of rice homebrew. I had a glass and a half, and there was much laughter to be had by all. The *chung* also makes understanding Nepali easier.

Remember the funky angled stairs up to my room from last year? Well, they are still here, and fun to negotiate after the *chung* . . . but I make it, on all fours! It isn't long before I fall asleep.

I then have the weirdest dream of my life. I dream that the dog begins to bark again at the rats going up the wall. After a few minutes, the hotel owner grabs the dog by the tail, takes it over to the stone fence and lops its head off with a machete. He puts the dog's head on the fence, exactly where they put the goat's head last year. Then he throws the dog's body on the ground and goes back to bed. A couple minutes later, the dog's head begins barking nonstop, and its body gets up and starts running around the yard, spraying blood all over like a lawn sprinkler.

I wake up and realize it was a bad dream. Thank God!

Until I hear it: *ruff, ruff, ruff . . . ruff, ruff, ruff!* Or is it a dream? I sit straight up and give my head a shake, then jump out of bed to look out my window, half-expecting to see the dog running around headless and spewing blood. I'm happy to see that it isn't, and then fully wake up to realize he's outside the same window, barking the rats again. By placing my ear next to the wall, I can hear them scurrying about. *Here I go again*, I think.

But then I say to myself, "Not this time!" I grab a few things, lock my room, and walk back to the Norling Hotel on the other side of the village. I ask the guys there how much for a room. The manager says 400 rupees, or $4 USD, but I can only stay for one night as they are booked solid the next day. I take it and go to sleep right away. Needless to say, I had a great night's sleep!

I sure hope there is Wi-Fi at basecamp, as I would really like to get word that I'm safe out to Sarmila, my ex-mother-in-common-law, Gloria (who I still call Mum), Purnima, and Ina from last year, as she said she would be following my progress closely.

In the morning, I go back to the Tashi Delek Hotel, take my things to my room and go for breakfast. I overhear a few people talking about not getting any sleep because of a dog barking all through the night. You'd think the hotel owner would take care of the rats in the wall so the poor dog could get some sleep.

After breakfast, we get an update on the condition of the monk from the helicopter crash. She is going to survive, which is great news, but she lost her legs. She has a long road to recovery ahead with her life-changing injuries. My prayers go out to her.

I go out for a walk and the morning is warm and sunny. As I walk, I see my little friend and her brother. She waves, then runs up to me and says "Namaste" in her sweet squeaky little voice. I reply, then give her and her brother a couple candies and take some selfies with them. Continuing on, I take some more pictures of local children with their little rosy cheeks and dirty tattered clothes.

At the far end of the village is a big stupa with a building (a holy building or temple of some sort, I think), surrounded by a stone wall with at least a hundred prayer wheels in it. I walk the complete circumference of the wall as I spin the wheels for good blessings from Buddha. I also take pictures of the many vibrant flowers here.

My little friend.

My little friend, her brother, and me.

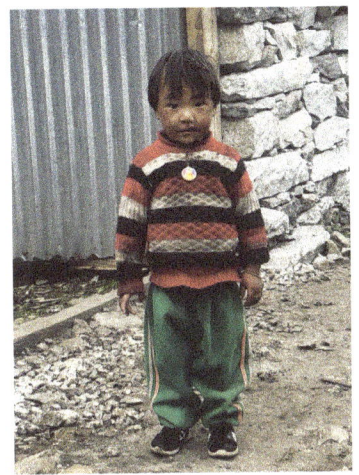

Boy of Samagaun.

Girl of Samagaun.

Intricate hand-chiselled tablets of Buddha.

Stone houses of Samagaun.

Here is an interesting curiosity for you. At the end of the village, I take a picture of the lush green valley with a plateau that transitions into steep slopes. From these slopes the Himalayan peaks stretch up to reach the heavens above, as the sun liberates morning dew into wisps of cloud that rise effortlessly into the abode of Shiva. It turned out to be a very nice picture but after I returned home and took a closer look at this photo, I could see an object of some sort in the air, about 100-150 feet above the plateau. I zoom in on the object, and I'll be damned if it doesn't look like a saucer.

Picture with UFO, left of centre.

To this day, I don't know what it was. The only aircraft that fly in this area are helicopters, as there are no airports here. The silhouette in the picture is not a helicopter; there were none flying for a few days, in the wake of the crash. There also aren't any drones around here. At home, I did some research and learned that there have been many UFO sightings in this valley throughout history, even before there were airplanes, according to locals.

I've arrived back at the hotel now, and there is no sign of the bag of toys for the kids. I hope it comes soon, because we are climbing to basecamp tomorrow. I spend the rest of the day just lazing around, relaxing and eating. We have heard there are many people waiting in Kathmandu who are going to be a few days late arriving, as there are no helicopter flights until the government opens the airspace again. Also, some people here are waiting for the flights to begin again to bring their gear bags. So these folks are stuck in Samagaun till their bags come.

I talk with the hotel owner, and he says he can let me stay tonight in one of the new rooms, at the end farthest away from the wall where the dog barks. It should be okay as these new walls are solid concrete. After dinner, I move my things to the new room. It's nice, and even has its own sit toilet and shower. The bed is comfortable with a thick foamy mattress, pillows, and quilts.

Monday, September 10, 2018

I had an amazing sleep last night and feel well rested. I get my gear ready as the porters outside prepared to haul our big gear bags and supplies to basecamp. Still no sign of the toys, which are coming by porter from Kathmandu and not helicopter. That's okay—I can give the toys to Lama Guru after we come back from the climb.

CHAPTER 16:
BACK TO BASECAMP

I'm now in basecamp, and what a slog that was! Even though I'm pre-acclimatized, it's still taxing to gain 1,000 metres (3,280 feet) in only four hours with a full pack. That's about three times what you should normally gain in a single day. We make awesome time, but once at basecamp I need to lie down in my tent for a half-hour as I'm feeling a little nauseous. After this brief rest I feel better and come out of my tent to check things out.

We had great weather on the way up, so I took some video and pictures of the view. We got to camp in the nick of time as it began to rain about ten minutes after arriving.

I'm now sitting in the common tent, considerably smaller than last year since there are only three of us. There is one table, four chairs, the heater, and all the munchies on the end of the table, including the ultimate climber's superfood, Penotti. The rest of today will be all about unpacking, making my tent cozy and just chillin'!

Our cook is the same person we had last year, and it's good to see him. This time he only has one helper, who is pleasant and accommodating. Basecamp is just as big and populated as it was last year. It is a veritable international city above the clouds and in the heavens. Looking down upon us from above is Shiva, the destroyer and creator, as well as Buddha.

Tuesday, September 11, 2018

Last night when I retired to my tent, it was so comfortable and no barking dog, bonus! I slept well, even though I woke up a couple times hearing the familiar sound of distant avalanches, which strangely enough I found comfort in. Welcome to Manaslu, Mountain of the Soul.

This morning my waking pulse is at 90 bpm and my oxygen was at 85%—not too bad I thought. I wake up with just a touch of headache in my forehead, but with some deep breathing, it soon dissipates.

Today, we will have our puja performed by Lama Guru and his assistant, a Lama-in-training, I believe. This time we have great weather for the ceremony, a little cloudy to start but becoming sunny. For the start of the puja, I am sitting cross-legged with my hands also crossed between my knees. When Lama begins to chant his prayers to the drum beat, a huge Himalayan bee lands on my hand, and just sits there for about five minutes, as if taking in the energy of the puja.

Entranced, I watch the bee on my hand as I fall into the experience of the ceremony. Strange that the bee would be up here—we are well above treeline and there aren't even any plants up here. Maybe it's a sign of good blessings from Buddha.

Years ago, when I was climbing Mt Cline in the Canadian Rockies with Jay Mills, on the glacier were hundreds if not thousands of ladybugs frozen in the snow. When the sun came up and warmed them, they reanimated and flew away. It was the strangest thing; even Jay with all of his mountain experience was surprised. We concluded that the ladybugs were carried there by upslope winds then froze in the snow. The bee visiting me now has probably also been carried here by upslope winds.

Once again, we drink sherpa soup during the puja, and afterwards we have some whisky with treats like chocolate, chips and cookies, feeling good from the alcohol. I still have the red and yellow blessing necklace around my neck from last year. Lama Guru places another one on me for this year.

His assistant speaks English, and through him I explained to Lama Guru that the toys still haven't arrived. He says not to burden myself with this.

When I go back to the village after the climb, he says I can take the toys to the Sama School, where most of the suffering children are. This is a great relief to me. Truly, getting the toys to the children is dearer to my soul than climbing this mountain.

After the puja, our whole team takes some pictures in our Satori shirts and the expedition banner in front of us. This time we have Lama Guru and his assistant in the photos with us. After this, I go to the Seven Summits camp next to us. Some of the sherpas there recognize me from last year, and we chat for a while. I ask them about the Wi-Fi situation in camp, and they tell me it will be set up in a couple days.

Manaslu basecamp looking towards Samagaun.

Manaslu basecamp at 4,800 metres (15,748 feet).

Lama Guru performing our puja.   *Lama Guru and his assistant during the puja.*

The 2018 Manaslu team. Left to right: Mingma Tenzi Sherpa, Pasdawa Sherpa, Radu Albu, François Matter, Lama Guru, Lama's assistant, Mingma Temba Sherpa, Gelzen and me.

I go back to our common tent, and a Seven Summits guide soon joins me. He's a big blond Dutch man named Arnold who now lives in Kathmandu and works as a professional guide—a very friendly and personable guy. He tells us that Samagaun is crammed full of people wanting to come up to basecamp, but they're still waiting for their gear to come from Kathmandu. Flights are shut down for now and no one knows when they will start again. I am thinking they will be shut down until the crash investigation is complete.

This year, the Chinese camp looks to be twice the size it was last year. There are many Chinese here with no climbing experience at all. Almost all have never even put on crampons before, according to Arnold. Sounds like it will be an entertaining year on Manaslu.

The only concerns we have with it being so busy is when the weather window opens up and everyone is pushing for the summit. How many traffic jams will there be, with people running out of oxygen because they are stuck in queue waiting for hours? I have to admit, I'm still a little nervous after last year, as now I know what I'm up against. I will just have to trust in my training, preparation and experience.

I've finished fitting my boots and setting up my harness; my basic gear is ready to go. Besides the bag of toys, I'm still waiting for my other bag of gear. It contains my rain gear and I'm sure I will need it, as Manaslu is the wettest mountain in the Himalayas.

Wednesday, September 12, 2018

I must have gone to bed too early last night, as I woke up a few times. It might have been the heavy rain that woke me, too, as it lasted all through the night and still continues.

My pulse/oxygen levels are great this morning, at 76 bpm and 86%. This time I also brought a collapsible Nalgene bottle to use as a pee bottle so I don't need to leave the tent at night, but last night I used it for drinking water. (It was okay as I hadn't used it for pee yet.)

All the water in camp is boiled, so it's hot when we draw it from the large insulated canteen. Last night I put the water in the collapsible bottle to cool and put it in my sleeping bag by my feet. It was so nice and warm in the bag, and in the morning I end up with cool drinking water. I don't mind leaving the tent at night to relieve myself, as it gives me a chance to view the amazing nighttime sky.

The falling rain has just turned to wet snow. The sherpas are considering going up to touch Upper Camp One, with this weather we might just be sitting tight for a while. Right now, some sherpas are up at Upper Camp One fixing the ropes and anchors. If they come back and say the bad weather is only down here in basecamp, then we will probably go for it.

* * *

We just got word and the weather at Upper Camp One is not too bad, just a little snow. Sounds like we are going. Of course, my umbrella is with my rain gear, which isn't here yet. Thankfully, Pasdawa has an extra umbrella that I can use.

It's a little wet getting to the crampon point, then sleety as we gain altitude. The crevasse field brings back lots of memories. We pass the spot where I became ill last year, but this time I'm feeling great, other than getting wet (despite the umbrella, as wind is driving the rain and sleet sideways). The glacier becomes steeper as we approach Lower Camp One, and the sleet is now a wet snow. It was also getting colder as we gained elevation.

We have been socked in with clouds all the way up, but maybe this is a good thing for me, as I'm feeling okay: I'm not overheating and I'm feeling good with the altitude. We pass a few tents at Lower Camp One and continue around and up to the shoulder. Finally, we reach Upper Camp One at 5,800 metres (19,024 feet).

Pasdawa shows me my tent and I hop in to rest for a bit out of the wind and snow. We stay here about twenty minutes and then start on our way back down to basecamp. On the way down, the weather really craps out on us. I am soaking wet and cold now as I try to stave off the wind with the umbrella. It helps, but my rain gear would have been much better. There's a chill in my chest and back, and I'm also getting a dry cough, which is a bad sign.

Both Pasdawa and I finally walk into basecamp and both of us are pretty happy to be back. I open up my tent and get in, then realize the vent flap on the top of my tent was leaking. My sleeping bag is wet, right where my lower back lays. I do what I can to dry it out, but it's still pretty damp.

Still no sign of my other gear bag. I really don't expect the porter to haul the bag up here in this weather; I sure wouldn't do it if it were me. The trail through the trees is a steep drainage path and would be slippery and muddy and very hazardous.

I haven't drunk enough water today and am a little dehydrated. I will need to get one of those CamelBak water bags that fit in your backpack with a tube to drink from as you climb. Because I always have to stop to drink, I then tend to not stop as often as I should. Because of this I don't feel that I'm drinking enough water.

So far I like climbing with Pasdawa. He is patient, relaxed, and not in a hurry. Back at basecamp, I am tired from fighting the cold and wet conditions, along with the altitude gain. I'm so tired, in fact, that I'm not very hungry and only have some hot soup for dinner.

I check my pulse/oxygen levels and I'm at 132 bpm and 59%, which is quite low. No wonder I'm so tired. It feels like I can't fill my lungs with air, almost as though my airway isn't completely opening. Maybe I'm beginning to come down with something.

My second attempt to climb Mt Manaslu might be quashed by illness. Not good! This time I forgot to get antivirals from the doctor; I only have antibiotics and Diamox with me.

Time for some sleep. I will put my large towel doubled up over the wet spot in my sleeping bag. *Subha ratri.*

Thursday, September 13, 2018

My sleep was disrupted last night due to this dry cough. François, Radu, and I are now sitting and chatting in the common tent. Radu is also becoming ill from getting wet and cold and has a bad cough. François, at sixty years old, is a machine and a very strong climber. Years of trail-running high up in the French Alps has paid off. He goes nonstop and is much better with the altitude than both Radu and me.

My throat is dry and sore, making it hard to swallow. Anything on the dry side triggers my gag reflex, and I can't get it down. For breakfast, we have hard boiled eggs and pancakes, but I only have the eggs and carbohydrate powder that I brought from home, mixed with water. In my past when bodybuilding I would use carbohydrate powder for energy as it is basic muscle fuel and more importantly complex carbohydrates are the primary fuel for your brain and a lack of this will usually result in lethargy and headaches.

Being at altitude wreaks havoc on the good bacteria in your digestive tract. Many of them die off due to lower oxygen levels, which is why many climbers suffer diarrhea in basecamp and higher. To offset this, I brought probiotics to replace these bacteria. Every day, I take a multi vitamin, probiotic, Diamox, gingko biloba, iron to help carry more oxygen in my blood, and aspirin to help with circulation by thinning the blood (your blood actually thickens at high altitudes). You can also take a couple tablespoons of vinegar every morning to help thin your blood. I have heard that some climbers take Viagra to help with altitude as it a vasodilator that opens up the vessels allowing blood to flow more easily.

Today's plan is to stay at basecamp, rest, and recover. Depending on the weather, tomorrow we will climb to Camp Two, then descend back to Upper Camp One to sleep the night. Radu tells Mingma Tenzi that he does not want to climb tomorrow and needs another day's rest to get rid of his illness; Mingma says for him to see how he is feeling tomorrow. I tell Radu that I will stand behind him and support his decision to not climb tomorrow if he has not recovered. I'm not in a hurry or trying to set any records; I only want to make the summit and back safely without feeling sick, or worse yet, dying.

I'm feeling so much better this morning and it's a beautiful sunny day, thank God. Mt Everest, as well as the rest of the Himalayas, are drier and colder mountains than Manaslu. At work I always said I'd rather work in -30°C and a blizzard than a cool day with rain. I hate the rain, because once I'm damp and chilled, I usually become sick.

Below us near the helipad is the Chinese camp. They have flashing, multicoloured lights strung all over their camp, and it looks like a discotheque. You can hear them down there partying, playing club music, hooting and hollering most of the night. We hear there are many people drinking down there, too . . . sounds like they are having a great time. The Chinese economy is doing very well, and it seems the Chinese have nothing but money to burn. I am zealous, I could help a lot of children with money to burn. Many Chinese people now come to Nepal to climb the peaks, as it's inexpensive for them. This is good, because Nepal needs the money. It's so sad that most of the money will not go to where it is needed. I recently read that China is also helping Nepal to rebuild after the 2015 earthquake. I think this is good as long as China does not want to take over Nepal.

At camp this morning, we're just trying to pass the time. Radu tells me that his illness is bacterial, but he is also sneezing lots. Maybe I'm wrong, but in my experience, while sneezes can be caused by bacteria, they are usually a viral and therefore a contagious illness. Either way, we are in a confined space in this common tent, and his illness has me concerned . . . especially when many people sneeze into their hand instead of their elbow, then handle things that everyone else touches. Just sayin'!

I brought a book to read, *Into Thin Air*, by John Krakauer which is about the 1996 disaster on Mt Everest. The movie *Everest* is based on this book. But I'm not much of a reader, except for technical books on topics I need to learn, like aircraft maintenance procedures or climbing and rescue techniques. I'm also not much into playing cards, so the long days in basecamp are excruciatingly boring for me. At the same time, we can't just run up the mountain. That's life in the Himalayas: hurry up and wait (and wait . . . and wait . . .)—then go, go, go!

It's official: the sherpas tell us we will not climb tomorrow. We will rest and go up the day after. That is good news, as we need more rest.

It's now 5:31 pm. I just checked my pulse/oxygen levels, and they are very good. I have a pulse of 79 bpm and blood oxygen saturation of 90%. That's awesome!

Friday, September 14, 2018

Once again, my sleep was off and on, so I'm feeling quite tired this morning. For the first time since arriving, it wasn't raining or snowing last night. The sky was crystal clear and stunning: all the stars gloriously twinkling, with the moon's warm glow and the Milky Way painted across the sky. It's also amazing to me how the snow-covered slopes of the mountains glow during the night, as though they have their own light. Last night we had pizza for dinner, if you want to call it that. With no tomato sauce, it was dry and difficult to swallow; I had to drown every bite with a mouthful of water to get it down.

The clouds and rain come back this morning, about a half-hour after I get up. While brushing my teeth, I notice that I've developed a gag reflex that

I don't normally have. I hope it goes away, but I begin to take Gravol daily, just in case.

I get into a conflict with the cook this morning. I'm often the first one up in the morning, and I usually go straight to the common tent for some coffee and to fire up the heater. Everyone else appreciates coming to a warm tent for breakfast, especially after spending the night in a cold, damp tent.

But this morning, the matches are too damp to light the heater. I go over to the cook's tent and ask him for a match.

The cooking tent is nice and warm, with the burners already lit and the water boiling . The cook refuses to give me a match, saying that soon the sun will be up and it will warm the tent. The truth is, every morning has begun with cloud cover, and even when the sky is clear, we are usually in the tent for a half-hour before the sun rises enough to even hit it.

I say to the cook, "Who pays for the propane—you or us, the customer?"

"You do," he answers.

I follow with, "Yes, this is our expedition and you work for us!"

I have to bite my tongue as I turn away and go back to my cold damp tent. It's funny, you know, because when I go back to the common tent to get a coffee—no more than five minutes later—the heater is now on.

This morning, my other bag with my rain gear shows up. The bag of stuffed animals arrives, too. I joke to François and Radu that we have stuffed animals to play with and keep us company during the nights. I open the bag in the common tent and notice that some of the stuffed animals are damp as well. I turn up the heater and pull out these various furbies and place them around inside the tent to dry. It looks like a petting zoo in here. It doesn't take long for them to dry and I place them back in the plastic bag and then in the gear bag.

Mingma Temba comes into the tent and tells us that the forecast is good for tomorrow; Radu says if not, he does not want to leave basecamp. I'm just happy all my bad-weather gear is here now, and I'm ready for any conditions.

It has been raining or snowing all day. Another boring day in camp, and I'm sitting by myself in the common tent. It's only 7:10 pm and the others have gone to bed. Tonight for dinner we started with yummy soup and popcorn, followed by rubbery yak meat, lumpy rice, and cold cooked veggies. I am not very impressed.

166

CHAPTER 17:

MINGMA'S AVALANCHE RESCUE

Saturday, September 15, 2018

I'm now in my tent at Upper Camp One. What a slog up the mountain that was with a fully loaded pack! I also brought some supplies to leave here like a couple bags of my food and some fleecewear. The other guys got here about 3/4 of an hour ahead of me, but that's all right as François is a climbing machine. This tells me I need to concentrate more on running once I get home.

When I first arrived, I had just a touch of headache in my forehead, but on the way up I was fine. The weather is good, with no precipitation, and I actually feel very good now that the headache is gone.

Upper Camp One at 5,800 metres (19,024 feet).

Above the clouds and in the heavens at Upper Camp One.

I'm also happy that I am able to get word out that I'm safe. Pema is talking with Lakpa back in Kathmandu, and he hands me the phone. Lakpa asks me how things are going, I tell him that things are going well, and ask if he can relay to Nabin at the hotel and his sister, Sarmila, that I am well. Lakpa says that he will which pleases me.

Sunday, September 16, 2018

Last night, I didn't sleep at all; a bad headache kept me up. Soon we will be leaving for Camp Two at 6,400 metres (20,992 feet). It should be exciting as we will be climbing through the icefall and seracs, Maybe the route will get more technical and I can have some fun. This is also the most dangerous area on the mountain.

* * *

It's 4:30 pm now and I'm in Camp Two. I still have a headache in my forehead and behind my eyes which goes away after some deep breathing. Thank God, this year I brought ten packs of dehydrated camp food in case I got bored or tired of the food here. You reconstitute the food by adding boiling water, and the big advantage is I can make it moist enough so I can swallow easily and not gag. Last night for dinner I had yummy spaghetti and meatballs, and have two more food packs I will leave in my tent at Upper Camp One for when we come back up. I am nicely surprised, as it was very tasty.

Monday, September 17, 2018

Climbing through the icefall is exciting and dangerous. Sometimes I can hear the house-sized blocks of ice shifting, cracking, and moving. Beyond the icefall are two steep ice cliffs that take the wind out of you. After each one I need to stop and rest. There are also two ladders lashed together to get up and over the last serac. At the top of the last ice cliff, a ladder spans a big crevasse—these are always good for getting the heart going as you cross them! Pasdawa stays just ahead of me and not once do I feel rushed by him.

We stay at Camp Two for the night. With the headache, once again I don't sleep a wink. It snows most of the night and stops at 2:30 am. It's eerily quiet, with the heavy snowfall absorbing background noise.

With the amount of snow, I'm concerned about avalanches. We're in a really bad place on the mountain if there were a slide, and this mountain has a notorious history of taking lives with it's deadly avalanches.

At 3:00 am, I go outside to pee and the sky is completely clear. I stand for a few minutes staring up at the stars. What a blessing, and a welcome distraction from my avalanche worries; not many people get to experience a view of the heavens like this. I feel like I'm at the center of the universe, with all the stars surrounding me. At this moment, I am so very grateful for all my blessings.

Headed through the dangerous and always shifting icefall to Camp Two.

Climber barely visible in the icefall.

Gargoyles in the icefall.

Massive ice daggers of at least fifty feet hanging above the icefall.

Ladders lashed together in the icefall.

View from Camp Two at 6,400 metres (20,992 feet).

Looking out from Camp Two.

The next day, the others are going to climb a couple hundred metres above Camp Two, then come back to Camp Two and stay for the night. Because I haven't slept well and still have the headache, I decide to go back to basecamp to get some sleep and recover. Pasdawa and I head back down, making really good time even though we passed through a couple traffic jams. We leave just after 9:00 am and arrive in basecamp at 11:20 am.

Before Pasdawa and I leave, Mingma Tenzi advises that I rest in basecamp for a couple days. But Radu says going back to basecamp now, without climbing a couple hundred metres and staying the night at Camp Two, will ruin my chance to make the summit.

"I am telling you!" he says.

"No, it won't," I say, laughing as we leave. Pasdawa and I had no idea what Radu was talking about. I won't do well with another sleepless night—I need sleep to recover.

A friend of mine who I met through the climbing community on Facebook, Mia Tucholke, is a mountain guide in Colorado. Before I left for Nepal, she reminded me to make wise decisions. Going back to basecamp is one of these wise decisions. Thank you, Mia!

Tuesday, September 18, 2018

I had a really good sleep last night and my pulse/oxygen levels were at 60 bpm and 91%. Not bad!

There isn't a cloud in the sky this morning. The weather is changing for the better, and I get the feeling we'll have a good weather window today. With any luck, the gods will bless us with a week and a half of nice weather.

The cook and his helper whose name I finally found out was Gelzen are playing Nepali music this morning, which I enjoy: the uplifting melodies and rhythms seem to invoke happiness.

Just after 10:00 am, the others come back to camp with quite the story to tell. They were climbing those extra couple hundred metres above Camp Two. A sherpa was freeing the fixed rope that had been buried in fresh snow, with his client right behind him. When the sherpa pulled up on the rope, it released the snow and caused an avalanche, burying the sherpa until only his hand and forearm were visible. The client was buried up to his neck.

Witnessing this, Mingma Temba Sherpa took off like a shot over to the buried climbers. He dug them both out with his bare hands, getting frostnip on a few fingertips from digging. Darn lucky that Mingma Temba was there to save them! The buried sherpa probably would have died within minutes, and seeing that no one would have been in the area for another day or two, his immobilized client likely wouldn't have made it through the night.

Now that the weather's nice, people are coming up to basecamp from Samagaun in droves. We can finally hear helicopters in the valley, so the Nepali government must have reopened the airspace. People's gear would now have made it to the village, too. The Chinese camp in particular is filling up with quickly.

If the weather window for the summit opens up soon, I suspect a lot of climbers will come down with altitude illness, since they haven't had enough time in basecamp for their bodies to acclimate. On top of that, many have no experience with climbing at all. They still need time to practise climbing with their crampons and jumar on the glacier beside the high end of basecamp where the training is taking place for the first timers. We are lucky to have gotten up here early, because when the weather window opens we will be ready to go.

Our dinner is dry overcooked chicken, soggy spring rolls, and dried-out, overcooked corn. I decide to eat a pack of my camping food instead; I'm so happy I brought these with me! Even Radu says he can't swallow the chicken—it's like dried out chicken jerky.

François seems to have picked up the same cough and sneezing that Radu is suffering from (although Radu insists it's only bacterial, "I am telling you!" he says), Radu sounds so funny with his plugged nose and his thick Romanian accent and they are both taking a cold medication that François brought with him. I'd better get away from these two petri dishes and go to sleep. *Subha ratri*!

Wednesday, September 19, 2018

Another beautiful morning in basecamp, considering I didn't sleep much.

Trying to avoid François and Radu's inevitable plague was no use: I'm now coming down with it. Just great!

I also had a strange dream last night that I was talking with Carol Burnett (a comedian that only us older kids will remember). I dreamt that I was telling her how I will help children through the youth society I'm starting up, and about the stuffed animals I brought for the children of Samagaun. She loved what I was doing. But when I reached into my pocket to pull out my phone to show her some pictures, my phone was missing. She then spent the next half-hour or so with me looking for my phone and she found it. I wake up and realize it was a dream, but a good one at that. I turn over and sleep for a little while more.

The sky is blue and it's warm today. As I'm enjoying the sun, a parasail launches from just above upper basecamp. Mingma Tenzi says that the flyer is the sherpa who parasailed off of Mt Everest. This sherpa plans to parasail down with an American woman from Manaslu's summit. We watch as he soars on the breeze then circles us a couple times. He flies into the valley below and comes back up towards us on the warm upslope winds.

He flies over the glacier beside us at the slope of Nadi Chuli Peak. But then, he loses control, spiralling down until he crashes into the glacier and out of sight. A rescue team leaves from basecamp to go help him. From being a competitive skydiver in the past, I know that 95% of people don't survive a spiral dive. François, who has experience with parasailing, says this is true for parasailing, too, and that the sherpa is more than likely dead.

It's been about an hour or two since the team went to help. They are not back yet. The mood in camp is sombre, as we fear the sherpa didn't survive.

It's just after 4:00 pm now. It's been another long, relaxing day in basecamp, but sad as we await word on the sherpa who crashed. We'll spend tomorrow here, and on the next day we'll be on our way back up to UpperCamp One.

We get word back on the sherpa who crashed his parasail: he is alive but injured. We hear that he was able to get himself to Samagaun and is on his way by helicopter to Kathmandu. We're relieved to hear this great news; he is very lucky to be alive.

There is still no Wi-Fi set up here. I still haven't been able to get word back to Canada that I am well, which is concerning because I'm sure my friends have heard of the helicopter crash and avalanches, especially the one where Mingma Temba dug out and rescued the two men.

Every night, I hear avalanches rumbling and thundering down the slopes high on the mountain, and every so often you can feel the mountain shudder under your feet. These larger slides happen during snowfall when the mountain is clouded in. You can't see the avalanche, but you know it's a big one. Sometimes you stick your head outside and watch, in case you need to prepare for impact.

Today has been unusually warm. I leave my phone in my tent, and it overheats, going into emergency shut-down. No wonder—it's easily over 45°C in my tent. I put my phone in some snow to cool it down. The heat is also loosening rock frozen into the slopes around basecamp, causing rock-falls. The trembling ground sometimes brings everyone out of the tents to see whats happening.

Like last year, there are many non-climbers out today trying their high-altitude boots and crampons for the first time. It's entertaining to watch them stumble all over the place, although maybe it's a side effect from all the partying.

One of these guys has his crampon fall off because it's really loose. He tries furiously to put it back on, but he's trying to put it on upside down, with the spikes facing his boot. You can tell he's frustrated. Finally, someone shows him how to put it on properly. When he gets going again, the other crampon falls off, and he still can't figure out how to put it on—so he hobbles away with the crampon in hand.

My cough is worsening and my chest is feeling warm and itchy. I think I'll start taking the antibiotics I brought, this seems to be turning into a full-blown chest infection. It feels like a really bad case of bronchitis, which I often got when I was still a smoker nine years ago.

It's now 4:37 pm and I'm resting in my tent. Soon it will be dinner time. I am so looking forward to dry, impossible-to-swallow food . . . *not!* Tonight is a good night for the bagged camping food I brought. As I'm eating my rehydrated food, I notice Radu isn't enjoying his dinner very much, and François keeps getting hungrier and hungrier. He'll eat anything set in front of him. I wonder if maybe he doesn't have some hyena in his DNA!

Thursday, September 20, 2018

This morning my pulse is 106 bpm and my oxygen is 86%. I didn't sleep at all last night—I was coughing all through the night trying clear my airways, almost to the point of vomiting. Because of this cough, I spent most of the night in a sitting position, coughing and unable to breathe.

I don't think going to Upper Camp One tomorrow is a good idea. I have a splitting headache, and I can hardly wait for the antibiotics and my immune system to kick in and get rid of this illness. I'm sure it isn't helping that I'm not eating much. All I'm able to eat for breakfast is some rice pudding, and I have to fight to keep it from coming back up. It looks and feels like maggots sliding down my throat.

About a half-hour ago, a helicopter came to rescue a Chinese climber. We haven't heard why; maybe he has altitude illness, or he could just be ill from the cold, damp conditions. It takes me back to last year, when I had to leave the same way. I hope he will be okay.

Poor François now appears to be quite sick as well. I laugh to myself because Radu, in his infinite wisdom, is trying to tell the doctor that he's not ill, and it's just from the dry air.

"I am telling you! You are not sick, it is only a dry cough." He says to Francois. Radu is a funny sort and very smart: no matter the topic, he has expertise and is not shy about interjecting his vast wisdom, starting with "I am telling you!" in his thick accent. He is also a great ice climber, so skilful that sometimes he doesn't need to use a rope or anchors on vertical ice. I wish I were that good! Even celebrated ice climber Will Gadd could learn much from Radu—I am telling you!

The other two go for a trek to crampon point; good, because they are not my cup of tea. I really miss the group from last year, especially my dear sister Ina. For lunch I have some of my bagged food, a beef stew that's very good and easy to swallow, too! Each bag is enough for two people, so it goes a long way for just me.

The cook is making lunch now, and it's been deep fried to the point of being overdone: crunchy yet oily at the same time. I am not quite sure what it is though. It looks like some sort of samosa's along with french fries, soup and chapati bread. I'm not sure of what kind of oil he uses or if it's even fresh, but to me it smells like burnt motor oil. Having worked in restaurants before, I know that this smell means you're overheating and burning the oil. Once that happens, you must throw it away and get new oil—not keep using it! If the cook asks why I'm not eating lunch, I will just tell him my stomach is upset. I am sure he is doing the best he can and I don't see any need to insult him.

I go to my tent and try to have a nap, but when the sun is out or the clouds are thin, it's way too warm in there. I spend the rest of the day relaxing, and once we get some good cloud cover I go back to my tent to sleep.

About twenty minutes later, my buddy Lakpa Sherpa from Pioneer Adventures comes by to say hello, as he's just arrived in basecamp. I mentioned Lakpa earlier: he's the sherpa who was in charge of the International Women's Expedition last year, who invited me for tea and cookies on my way up to basecamp.

I've kept in touch with Lakpa ever since. I was even thinking about climbing Mt Ama Dablam with him after this trip, but I couldn't afford it. Probably for the better, as I should just concentrate on Manaslu.

Lakpa invites me to his camp to discuss going to Mt Everest in 2021. I tell him I will come see him tomorrow.

For dinner, the cook makes delicious soup and what he calls American chop suey. It's undercooked egg noodles with mushroom sauce and a cold fried egg on top. I'm not able to eat it.

CHAPTER 18:
MANASLU COMES TO LIFE . . . EARTHQUAKE!

Friday September, 21, 2018

Well, I had a hell of a night! Besides feeling ill, I have a major chest infection and a nasty case of the "Khumbu Cough." This type of cough is named after the Khumbu Icefall on Mt Everest, caused by the extremely dry air. My throat and airways are very painful, feeling like shredded paper. My cough is nonstop, so much so that I'm choking and gagging, almost to the point of vomiting. It's also giving me a bad headache. My throat is so dry that I almost double over in pain whenever I'm able to swallow.

As I'm up that night coughing, I remembered where I've run into this type of cough before. It was back in the winter of 1993 when I refurbished and repainted airplanes and helicopters. That winter was extremely cold and dry. Before painting the aircraft, we would soak the hangar's floors and walls with water to keep the dust down. It also raised the humidity to help the paint cure. When spraying the paint, the air make-up unit removed the contaminated air by filtering the paint fumes and excess spray as well bringing fresh warm filtered air into the hanger. This process draws all the moisture from the air, also your skin and lungs, just like how washing your hands too often will dry them out.

We would get this dry, painful cough that we called "painter's cough." I found the only way to relieve it was to sit in the bathroom at home after work with the door shut and the shower running steaming hot. As I inhaled, the steam moistened my airways and throat. Last year here in camp, everyone was using a similar method, the steam baths with Tiger Balm oil that I mentioned at the end of Chapter 10.

I try one of these steam baths, and it's very relieving. I wish I had done this sooner! It's so soothing, in fact, that I plan to do it at least twice a day. When I get back home, I'll ask a pharmacist if there are any inhalable products that moisten airways; something like that would be helpful up here.

When I quit smoking cigarettes nine years ago, I had all sorts of sinus problems, including dry sinuses. I used an inhalable moisturizing spray called Rhinaris, and now I think this would be beneficial. With all this coughing and weight loss, my abs are going to be ripped!

Before going to bed last night, I took a nighttime cold-and-flu tablet, which suppressed the cough enough for me to get at least an hour or two of sleep. The rest of the night, I listened to heavy wet snow landing on my tent and the nonstop rumbling of avalanches. When I emerge from my tent this morning, it looks like a Bing Crosby White Christmas with all the fresh snow.

Basecamp after the snowfall.

Basecamp, looking more Arctic than Himalayan.

Three to four inches of wet snow cover everything, and the tents all look like little igloos. It could be a bad day avalanche wise for anyone at the upper camps. As well as the increased difficulty climbing in fresh deep snow known as post holing.

* * *

It's late afternoon now. The snow is still coming down like the dickens and it's not supposed to stop until tomorrow. I have to be careful when I go visit Lakpa, since it's so slippery on all the rock.

His camp is set up comfortably, and I spend time chatting with Lakpa and all his sherpas. A couple of them are quite famous, and they are heroes, as members of the newly formed rescue team stationed on Mt Everest. They have risked their lives many times already to save people. I am honoured just to be in their company. If you would like to see these heroes in action there is a very good video on YouTube called, Everest Sherpas: 'They're not heroes. They're rock stars'. I cannot give enough of a shout out for these brave men who regularly risk their lives to save others in the worst of conditions.

Then the boys break out the Johnnie Walker special reserve and some Chinese food. Big Arnold from the Netherlands joins us too. I only have a Coke, though—no alcohol for me while I'm recovering, and I don't want to risk slipping and hurting myself. I stay about an hour, then go carefully back to my camp.

Once back at my camp, I do another steam bath and ask Gelzen to add six drops of Tiger Balm oil. It's strong enough to make my eyes burn, but it feels so good to breathe in. I then eat one of my bags of freeze-dried food: lasagna with meat sauce, yum! I'm famished and gobble down the whole bag full. I guess this means I will have no room for dinner—oh, well.

It's still snowing like crazy outside, and the avalanches are a nonstop rumbling. Not to worry, though. Basecamp is relatively safe, unless there's an earthquake like the one in 2015 that took out the Mt Everest basecamp by triggering a massive avalanche.

There's a guy from Seven Summits at Lakpa's camp setting up the Wi-Fi. He's going around collecting money and giving people the password. It costs thirty dollars for forty-eight hours. I don't have any cash on me right now, so I will have to check on this later.

Saturday, September 22, 2018

I leave my tent this morning at 1:00 am—I urgently need to pee (my back teeth must be floating!). Beside my tent is a bit of a drop-off were no one travels. As I start to relieve myself, a thunderous *BOOM* echoes throughout the valley below., loud enough to make me duck.

Right after the boom, it feels like the entire mountain jumps up a few inches and drops back down. Before I know it, there's rock, snow, and ice falling all around basecamp. I figured it was an avalanche big enough to shake the mountain, causing the boom and triggering other avalanches, so I think nothing of it and go back to my tent and crawl back into my sleeping bag.

I got maybe three hours of sleep last night due to my cough, and from about 2:00 am onward it's really hard for me to breathe. I'm wheezing and it feels like I have a marble stuck in my throat.

When I sit up in my bag, I hit myself hard in the chest a few times to try and clear the blockage. I feel that something changes in my chest, so I go outside and cough as hard as I can while bent over. The chunk of phlegm I hack up looks like a dead squashed mouse on the rock. Thank God I'm at least able to breathe now!

This is a good sign that the steam baths and the antibiotics are working, and also confirms to me that I have a bad bronchial infection. I used to get these when I smoked (a pack a day for thirty years, although I've been cigarette-free for nine years now). After clearing my airway, it's almost time to get up. My pulse this morning is 96 bpm and my oxygen is 88%.

Radu and François also heard the boom and felt the ground shake last night as it woke them both up. Radu thinks it was an earthquake. I disagree—no way that was an earthquake, you wouldn't hear a boom first. Then I say that it was just a huge avalanche, one big enough to shake the ground when it landed. We argue over this for what must be a half-hour.

Then our cook comes in and tells me he has heard me coughing at night and there is a doctor I can visit at the Seven Summits camp right next to us, so after breakfast I will go see him. Maybe he can give me more antibiotics, as I only have two of my six Azithromycin 250s left. I think this is the team that big Arnold is with as a guide. I then walk up to Seven Summits camp and inquire about the doctor in their common tent and sure enough Arnold is there and directs me to the dome tent where the doctor is.

My prognosis is good. The doctor, Max, says my heart sounds good and my lungs are clear, but I have bronchitis. He thinks I'm on the mend, which is good to hear from a doctor. Funny thing is that he doesn't seem like a doctor, so when he takes my blood pressure, it doesn't shoot up like it normally does! He gives me more Azithromycin, twice the strength of the ones I have, and Strepsils for my throat.

Then he shares a little-known secret about climbing in the Himalayas: always wear a bandana over your mouth, breathing in through your mouth and out through your nose. The rock up here contains microscopic crystals that are like shards of glass. He shows me by shining a light on a small rock from outside; I can see the crystals sparkling. Many climbers get a cough and infections as these crystals act like tiny razor blades, cutting up your throat and airways.

He tells me to always breathe through the bandana and to keep doing the steam baths. The best part is that all this medical care, as well as the medication, doesn't cost me anything, and it sounds like I will be okay.

While I was waiting to see the doctor, I sat and chatted with the sherpas in the cook tent. They figure it will be a few days before anyone goes to any of the higher camps because the avalanche hazards are severe thanks to all the recent snow. But as soon as I get back to camp, Radu says that he thinks we will be climbing the mountain tomorrow!

"Not me," I say, as I need a day or two to recover from this infection. I can't wait until it's cleared up. My chest right behind my sternum feels like it's on fire.

Lunch is actually pretty good: chorizo sausage, boiled potatoes with melted cheese, and pomegranate.

Remember the climber from China I mentioned who needed a helicopter rescue? Well, today we find out why he had to leave. Get a load of this! According to a sherpa who witnessed it, the climber was up in Camp Two and ran into an ex-business partner whom he owed money to.

The ex-partner wants his money back, of course, but the climber says he doesn't have any money with him—so the other guy tries to take his climbing gear. A fight breaks out between the two of them. They are in a kung fu fight, wearing crampons as they kick each other! The guy who owed the money is bleeding like a stuck pig from the crampon wounds, and has to be flown to a hospital in Kathmandu for treatment. Too funny . . . there really is big trouble in little China!

Pema Sherpa stops by for a visit, and he lets me use his phone to call Lakpa in Kathmandu. The sherpas have Nepali SIM cards and are able to get cell service up here. I ask Lakpa to send a message to "Mum" (my ex-mother-in-common-law, Gloria) to let her know I'm okay. Because Lakpa will message Mum from Kathmandu, I now have no need for the Wi-Fi in camp.

The wet sticky snow is falling again. Wow—we are getting a lot more snow than last year! I'm going a little stir-crazy stuck here in camp, and I'm even getting restless leg syndrome. Very annoying! On the other hand, I need this time in camp to recover. The definition of bittersweet, I think!

This time spent in camp with the infection has turned out to be a suffer-fest and a real test of my fortitude. On occasion, I have internal arguments with myself. One side of me wants to give up, making up any excuse to leave and go home. My other side is doing all it can to remind me that I need to stay, not for myself but for the sake of helping suffering children, both here and at home in Alberta. This is so much bigger and more important than myself. I need to stay strong and stick this out.

Pema says that he, Taro Yamagata, and their group will start heading up the day after tomorrow. This would be good timing, as I should be feeling much healthier by then. I have been coughing up all the blockages in my airways and I am able to sleep more easily.

After dinner, Mingma Temba Sherpa comes into the dining tent and tells us they've received the weather forecast from the Indian Navy. According to them, our perfect weather window will be September 28, as it will be clear, warm and calm. The 27th will be cloudy with possible snow, and the 29th will be clear but windy and cold, which increases the risk of exhaustion and frostbite at altitude, especially in the death zone above 8,000 metres! The Indian Navy's forecasts are accurate, so summit day is the 28th—six days from now.

I'm hoping I can recover enough by then and get some strength back. But if I need one more day to recover and climb to the summit on the 29th, there should also be fewer people going to the summit and therefore less time spent in traffic jams. Still, the winds that day are expected to be twenty-five to thirty-five kilometres per hour at the summit. I guess I will know for sure in a couple days. I will need to stop taking the antibiotics when we start climbing, as they have been wreaking havoc with my guts and causing diarrhea. Not a great thing to have while climbing for obvious reasons—and it's also really dehydrating which drastically increases the chance of getting altitude illness.

Another issue is cropping up. A couple years ago, I strained some ligaments in my left shoulder. I went to physiotherapy at the time rehabbed it and recovered well, but I think I slept in a bad position, and the cold, damp conditions are affecting it, too. When I get back home, I will have to see my amazing physiotherapist, Andreanna. For now, I will use an anti-inflammatory cream and hope it doesn't become worse.

Sunday, September 23, 2018

I finally have a good night's sleep without much coughing. My waking pulse is 92 and my oxygen is at 88%.

We got some heavy snow last night, at least six inches' worth, making my tent sag. Because of the insulating factor of the snow, my tent stays pretty warm inside. There's a puddle of water at the foot of my tent, and the outside of my sleeping bag was also wet. I will need to pull my bag out of the tent and dry it out today, as well as removing the water from my tent.

The sky is crystal-clear this morning and the sunshine feels just wonderful. I have only been up about an hour, and two helicopters have already rescued people from basecamp, for reasons yet unknown.

The avalanches this morning have been nonstop since the sun rose and started to warm the slopes. Many of these slides come off Naike Peak that overlooks basecamp, sometimes making all of basecamp rumble. I bet they would register on a seismograph!

Pema, Taro, and their group are camped lower than us, about two hundred feet away from where the snow and ice is landing and accumulating. It's probably very exciting for them. I am not sure I would want to be camped where they are. To close to the avalanche zone for my liking.

I'm definitely feeling better this morning, and pray that I'll be healthy enough to go tomorrow. I make my gear list for tomorrow to make sure I don't leave anything behind, like my umbrella or sunscreen. I also just finished doing another potent steam bath; I love how soothing it is to inhale. I am now doing two strong steam baths per day. I hope that I can regain enough strength to make it up this mountain and back down safely.

Lunch was beans and sliced apple, and my stomach seems to be handling it just fine. I also treated myself to a mini Toblerone and a Snickers bar. For dinner, I think I'll have one of my bags of dehydrated food to make sure I have enough calories in me for tomorrow's climb.

Uh oh! Now I feel like a heel! We just learned more about that boom a few days ago. It was an earthquake! The epicentre was at the base of Mt Manaslu, magnitude 5.4 on the Richter scale. I'm eating crow now!

I feel terrible and apologize to Radu. I also chuckle, though, because he says to me, "I was trying to tell you, I am telling you!"

CHAPTER 19:
MY FURRY LITTLE FRIEND

Monday, September 24, 2018

I had a good sleep last night and today is the big day. I feel good and the weather is perfect.

I'm already packed, I've finished having breakfast, and I'm ready to go. This time I won't forget the sunscreen! I'm bringing my umbrella for sun as much as rain. Our packs are very heavy, as we have all we need for six days on the mountain.

All goes well up to the crampon point, and I'm feeling strong. Word of the weather window on the 28th has gotten out to everyone, and an endless stream of climbers is leaving basecamp. It sounds like there will be around three hundred climbers plus their sherpas all pushing to the summit on the same day. Because of this, we will be leaving for the summit the night before to ensure we arrive just before sunrise.

Ready to leave basecamp for Camps One, Two, Three, and Four, then the summit.

Headed out with full pack on.

Leaving the crampon point and rockin' the umbrella to avoid heatstroke.

Pasdawa and I rest at the crampon point for fifteen minutes to drink, have a snack, and put on our harnesses and crampons, then we start up the Manaslu glacier. Once we get over the steep toe of the glacier, the clouds come in and it stays nice and cool through the crevasse field as we weave around or jump over the voids.

No lack of excitement as we climb up to lower Camp One, with avalanches roaring down Manaslu's lower slopes onto the glacier to our left. We're travelling up the middle of the glacier, so we have no cause for concern. We can feel the vibrations of these monsters through the ice as they come rocketing down the slopes and impact with the glacier. I find that this is actually very exciting.

Up past Lower Camp One, we are now in the clouds themselves. It begins to snow ever so lightly, the fluffy flakes tickling my cheeks as they land and melt. The temperature is also dropping considerably. As we climb around the shoulder that enters into Upper Camp One at 5,800 metres (19,024 feet). I notice that the different layers of snow and ice on the shoulder to our left resemble a gigantic vanilla layer cake with white icing that has had a piece cut out exposing the layers inside.

Wow, Upper Camp One is full of people! There's even a stray dog up here. It looks like one of the dogs from Samagaun, except this little girl doggy has some black patches mixed in with her white fur.

My first thought is, *Oh no! It had better not be barking all night.* Then I realize that there wouldn't be any rats up here for it to bark at, so things should be nice and peaceful. Someone on the trail must have fed this dog, and she followed them up.

I settle into my tent. Someone's been in here and gone through the things that I left last week. The two bags of my dehydrated food have been stolen. This angers me, but what can I do? Often, climbers who are too cheap to pay for sherpa or guide support will climb a day early and hijack other people's tents to sleep in.

In the future I will lock my stuff up with a small padlock on the tent. I wouldn't have left this unlocked, but I was told theft is not very common up here. Thank God I have another bag of food with me, and can use it for two dinners. Tonight I will have what the sherpas have brought up for dinner. (Noodles, as it turns out.)

Now it has become very cold, I got a bit of a chill between here and lower camp one. So I put on my down suit as soon as I arrive. So nice and warm!

Now it's time to get some sleep. It's getting dark and a breeze picks up bringing the temperature down even more. I see the little dog is shivering, trying to get into people's tents. The dog is small and doesn't have thick fur. Of course I feel sorry for her. I give a quick whistle. The dog hears me and runs over. I still had some noodles left from dinner, so I put them on top of a Ziplock bag and give them to the dog to eat. She gobbles it up quickly.

I get into my tent and let the dog in, too. I would feel horrible if she got frostbite or froze to death. I use some of my extra clothing to make the dog a bed beside me. I fall asleep and my furry little friend does the same.

Tuesday, September 25, 2018

I had a very good sleep, as I was almost too warm in the down suit. I think my little friend also slept well.

I let her out of the tent, have some tea, granola bars, and an apple. The sherpas make tea first thing in the morning. They bring it to your tent, as well as corn flakes with warm milk, but I just have the tea. I'm not sure where the dog has gone. She's probably in another tent having breakfast.

After getting home, I learn that the dog found her way to Camp Two, meaning she crossed a ladder and climbed through an icefall. Or else someone carried her, which I would not be happy with—it would mean the dog is essentially stranded there now. Thankfully, another climber found the dog on the way down and got her safely back to Samagaun.

I get my things ready to go. Mingma Tenzi says since I'm a little slower than François and Radu, I should leave first and they will catch up with me. I start climbing towards the icefall.

It's a beautiful morning, almost too warm for climbing. I'm only wearing my softshell pants, a T-shirt under my light jacket, and the big, warm boots. I stop often to drink and cool down, as I'm sweating easily in the intense sun. Every once in a while, I grab a handful of snow and rub it on my neck and face. This feels so good and refreshing.

Fifteen minutes after leaving Upper Camp One, I am in the icefall again. I climb through and around the seracs and ice gargoyles, which are big, stand-alone blocks of ice shaped by the movement of the glacier. Some of the them are huge and stand precariously! Ice gargoyles are so surreal and cool to see, but they can shift and fall on you at any time without warning, so it's imperative to get through the icefall quickly.

Beyond the icefall, I'm on a straight, steep section. It's busy, so I'm basically in a conga line of climbers. Everyone is moving steadily, hopefully anyone who needs to stop stomps out a flat spot in the soft snow beside the hard pack trail, out of the way of others on the other side of the rope. The proper, safe thing to do if you are resting beside the trail is stay clipped to the rope until someone wants to pass. You unclip momentarily to let them by, but only if you're in a secure spot. One slip while you're unclipped would be a life-ender as you take the grand tour all the way down the mountain.

At the top of this steep section is a big crevasse that can only be crossed by walking over a ladder laid across it. Most climbers place their feet between two rungs, with their toe points on one rung and their heel points on the other. I have a different method, since I have experience on my toes from ice climbing. I walk across smoothly on my toes, one rung per boot.

Ladder crossing over the crevasse.

The others catch up to me as I'm resting on the side and then pass. François says he thought he would have caught up to me much sooner, and that I'm climbing well. This is nice to hear from him, and I thank him.

We climb through some even steeper sections. I don't find them that difficult, but some others are having trouble ascending the rope. They are leaning back on the rope, rather than getting their weight over their feet and leaning into the ice while using their legs to climb. Learning to ice climb would help them greatly, as they are wasting so much energy. Sometimes when out in the Rockies ice climbing I will warm up by climbing on belay without my ice axes. Just working on good foot work to climb while I lean into the near vertical ice. Just like rock climbing it's mostly about the feet and legs.

About eight hours later, I arrive in Camp Two at 6,400 metres (20,992 feet). My chest infection has taken its toll. I am moving much more slowly than the others, but my only concern is to make it to camp before dinnertime.

I need to give credit where it's due: Pasdawa's patience with me is nothing short of heroic. He must be dying of boredom—his sherpa brothers are already relaxing in camp, while he is still climbing and waiting for me.

Once in camp, I settle into my tent then finish my bag of food, yummy spaghetti and meat sauce. After eating, I hop into my sleeping bag and fall to sleep.

Wednesday, September 26, 2018

Traffic jam through the notch.

I wake up from another good sleep to a nice morning. I get my things ready, eat a little breakfast with my tea, then leave for Camp Three.

The upper part of the glacier is probably the shortest section on the mountain, but still takes my breath away. We run into a bit of a traffic jam climbing through a narrow notch just before the last ice cliff. This seems like the most technical spot on the whole mountain, and inexperienced climbers are causing the jam. Some even need the sherpas to pull them up on a rope. This is a technique called short roping which is also used in rescue.

As we get closer to Camp Three, this section becomes fairly steep. Two German nurses join me for about twenty minutes. The ladies are concerned about my dry cough; I explain that I'm recovering from a bronchial infection, and this plus the extremely dry air is causing the cough. I tell them that I saw the doctor and he said that I was on the mend and good to go. One of the nurses tells me that I should go back down to basecamp until I am better. I tell her I will be fine, but she keeps insisting I turn back. I get tired of hearing her, so I stop to let them get well ahead of me.

All goes well, and it takes me just over four hours to get to Camp Three at 6,800 metres (22,304 feet). An hour before arriving, Pasdawa went ahead of me to make sure everything at the camp was ready for us. It's a little windy when I get there. Camp Three experiences converging winds because it sits on a flat saddle just below the col of the mountain, between the summit and the east pinnacle. A wall of ice mushrooms—ice formations shaped by the wind—extends above the camp.

The camp is really big, with four sections: upper, lower, and on either side of the trail. At first, I have trouble finding our spot. Exhausted, I wander around for about fifteen minutes, and just as I'm getting frustrated, I finally spot Mingma Tenzi. I'm so happy to see him and our camp! I crawl into my tent immediately, lying down to relax. I take some deep breaths to get more oxygen into my system, as I'm feeling depleted. After about ten minutes of this, I'm feeling much better.

I don't see much of François or Radu, but since I'm usually at least an hour behind them, they're probably in their shared tent, resting in their sleeping bags. As the only one on our team with Adventure 14 Peaks, I get my own tent, except for Camp Four, where I will share a tent with Pasdawa and Mingma Temba.

On my way to Camp Three.

The wall of ice mushrooms above Camp Three at 6,800 metres (22,304 feet).

As I'm relaxing in my tent, Mingma Temba brings me some black tea. I can hardly believe my eyes—he is barefoot in the snow! Talk about a tough cookie. It is freaking cold up here, too, at least -20°C. The sherpas really are superhuman!

For dinner, I finish my bag of spaghetti with yummy meat sauce, then I ask Pasdawa if he could fill my collapsible water bag with hot water. When he brings it back, I place it in on my chest—it feels so good, since my chest is still sore from the infection. Once my chest is nice and warm, I put the water bag at my feet to keep them toasty as I fall to sleep.

In the upper camps above basecamp, there are designated holes dug in the snow for use as washrooms. For privacy, these latrines have snow walls built on the camp side, but the other side is open to the wind—and Camp Three is usually very windy!

At about 3:00 am, I need to use the facilities. It's -32°C, according to the thermometer on my altitude watch and this is the temperature inside my tent. Outside with the windchill, it's probably closer to -40 to -45°C. Thankfully, the down suits undo completely at the back so you can let it all hang out and do your thing. Let me tell you . . . you don't let it hang out very long in these conditions, boy! When the climbers are all off the mountain and the sherpas take the camps apart, they also scoop up the (now frozen) human waste and put it in thick plastic bags for removal.

Before I hop back into the tent, I look up at the sky. It's crystal clear and oh, my God, I've never seen anything so beautiful, yet at the same time making me feel so minuscule. I feel like I'm in the heavens and the house of God, which is ironic as my last name means "the house of God." My first name means "beloved," so together they mean I am beloved in the house of God. But these are the Himalayas, so I am beloved in the abode of Shiva, I hope.

I now feel that everything will be okay. It's only fitting, this view of the sky tonight as my father was an astronomy lover. He and I would sometimes sit outside in the backyard at night, looking at the constellations with my small telescope. He even taught me how to make my own star charts and take astrological readings. He would love to see this sky tonight, and I hope he's doing so through me.

Thursday, September 27, 2018

I wake up after a good sleep, feeling well rested and stronger today. Good thing, too, because getting to Camp Four will probably take me eight to ten hours, as the air will be very thin and I will not be on supplemental oxygen. I know I will be slower arriving at camp, but I want to make sure that I maintain a pace where I feel healthy and comfortable. I don't want altitude illness to force me to turn around. If at all possible it would be nice to finish this climb without any headache so I can actually savour this journey rather than just surviving, so I am in no hurry.

I have a bit of a cough left and still feel the weakening effects of the chest infection. The climb has become a balancing act for me: I must travel fast enough to get to the camps in decent time, but also slow enough for my body to maintain its ability to deal with the lack of oxygen. I also have to deal with the cold and the exertion in my weakened state.

Pasdawa and I leave for Camp Four. The route takes us through the top section of the glacier, weaves around a few seracs, and up some short steep

sections. It's slow-going here for me, and Pasdawa goes ahead to make sure everything is set up for us in camp. At this point, I am taking a step then a breath or two, a step then a breath or two (or three), and stopping whenever I begin to feel my heartbeat in my head, which is often.

I'm now on the last section before Camp Four. It's an extremely steep traverse with high exposure to falling. If you fell here without being clipped to the rope, for sure you would take the grand tour . . . ending with your death thousands of metres below. Staying clipped into the rope here is critical!

En route to Camp Four at 7,450 metres (24,436 feet).

CHAPTER 20:
ASCENSION INTO THE ABODE OF SHIVA

Thursday, September 27/Friday, September 28, 2018

I arrive at Camp Four (7,450 metres or 24,436 feet), and I feel really good as my pacing worked well. It takes me just over eight hours to get here.

I have some noodles, a sliced apple, and some chocolate with Pasdawa and Mingma Temba before getting my things ready to go to the summit. At about 8:00 pm we go to sleep. I wake up just after 11:00 pm, nicely surprised that I slept deeply and well.

We get our gear on, and because it's very cold outside I sit in the tent door with my boots sticking out to put on my crampons. Under my boots, I'm wearing my thick summit socks for the first time, and electric heated socks. As I'm getting ready to leave, I can hear Radu in his tent saying, "No, I don't want to get up, I want to sleep some more." I get a laugh out of this—he sounds like a teenager having to get up for school in the morning.

Pasdawa sets up my oxygen for me, we don and turn on our headlamps, then head out into the night. Our headlamps slicing their bright luminescent beams through the darkness of the peaceful Himalayan night while the stars and the Milky Way shine their blessings upon us. The stars are so close I can almost reach out and touch them. Up here with much less atmosphere it seems as though you can see them in 3D. I feel blessed to be here experiencing this, and feel the presence of my mom and dad's spirits with me. This transports me to a place of immense peace and happiness, which I feel to the depths of my core. Wow, this feels so spiritual. About ten minutes later, I catch up with Mingma Tenzi and Pasdawa. Pasdawa turns my oxygen up a little. A wave of warmth flushes down through my body to my toes. My boots are warm now, so I turn the heat level down on my socks (with my phone, as they are connected with Bluetooth).

A funny thing: all I can think about now is a frosty, ice-cold bottle of Coke, poured over ice, in a frosty, ice-cold mug. For so long while climbing, I keep imagining pouring the Coke over the ice and taking a drink, with the foam sprinkling my nose as the sparkling bubbles pop.

Now the route is becoming very steep as we climb the final pyramid slope to the summit. Again, we are basically in a conga line into the void of space. Above us, a string of headlamps illuminate the near vertical ice and snow in hues of blue and white, rising into the starlit heavens above. Below and back to Camp Four, three converging lines culminate from camp into a single column of blazing light that follows our footsteps. We are ascending to the pinnacle of the Mountain of the Soul and into the abode of Shiva.

Pasdawa passes some slower climbers, then gestures for me to follow him. On either side of the well-packed trail, the snow is two to three feet deep. Passing the slower climbers is exhausting for me, since I am not a sherpa or a yeti, and soon after I'm forced to stop to the side of the trail and rest. I overstepped my pace when passing. The slower climbers, with their steady pace, then pass me as I rest. Pasdawa soon learns that it's not good for me to run past slower climbers in deep snow.

After a series of switchbacks, I start to feel ill. I'm very cold and I slow down drastically. I have to stop, as I'm feeling hypoxic. Pasdawa checks my oxygen and finds that my tank is empty. My oxygen should have lasted longer than this; it must not have been full when he gave it to me. Pasdawa replaces my oxygen tank and I feel great again with that familiar flush of warmth throughout my body. I actually feel a warm fuzzy rush of warmth that travels out to the tips of my extremities.

We are now climbing above 8000 metres into the death zone. Here, there isn't enough oxygen available to support life, so our bodies are slowly beginning to die now. It is just a matter of time. With the lack of calories and extreme exertion at altitude our bodies are consuming muscle mass for energy. Three quarters of the way to the summit, once again I start to feel slow and ill, needing to remove my mask to get any air at all. I ask Pasdawa to check my oxygen again, and my regulator is frozen solid. Pasdawa changes it out and all is well again, or so I think. Pasdawa is still wearing his mask, so I don't realize that he's climbing without oxygen. I assumed he gave me a spare regulator, but he doesn't have a spare, and actually gave me his.

At the fore summit (8,100 metres, 26,568 feet) after climbing through the night.

I must be dehydrated, because all I can think about is that frosty cold mug of Coke. Concentrating on this seems to keep me going as I trudge towards the summit, placing one foot after the other.

We make it to the fore summit at 8,100 metres (26,568 feet). It's a plateau. We are just under the true summit as the sun rises, casting hues of orange alpenglow upon us. We see the others on our team and everyone is happy to be up here. The rising sun is wonderfully warm, without a single cloud in the sky.

I pull my phone out to take a picture, but the last time I used it I put it in a colder pocket, and it froze. I'm greatly disappointed—I miss taking the picture of a lifetime, when the shadow of Manaslu casts a perfect pyramid of darkness off into the distance and on the rest of the Himalayas below. The view with the sun coming up is absolutely spectacular.

Pasdawa then confesses to me that he's suffering from a severe headache, he's been climbing without oxygen since my regulator froze. I feel terrible, now realizing that he sacrificed his own well-being for my safety. I offer my oxygen to him, at least to clear up his headache, but he refuses to take it. We find Pema Sherpa, who helps Pasdawa get his oxygen fixed. The sparkle then comes back to his eyes.

I turn my electric socks back up, as we're now waiting in a queue to get to the true summit at 8,163 metres (26,761 feet). The true summit only has enough room for two or three people. Most people stop just a few feet before the very top because you risk falling off the other side, as one climber found out a few years ago. He was unclipped to get the few extra feet to the very top, while taking a picture, slipped, and took the grand tour. It is so busy that Pasdawa and I wait for almost four hours before we are able to celebrate at the summit and take pictures.

The long wait to take pictures at the true summit.

My partner and guide, Pasdawa Sherpa.

Almost at the true summit, looking down upon the Himalayas.

Looking down on the east pinnacle, Birendra Lake and Samagaun.

Summit selfie.

*The Canadian Rockies Youth Society flag
proudly flown from its first 8,000-m summit!*

Pasdawa and me at 8,163 metres on the summit of Mt Manaslu.
You can see by Pasdawa's tan that the sun at altitude is very intense!

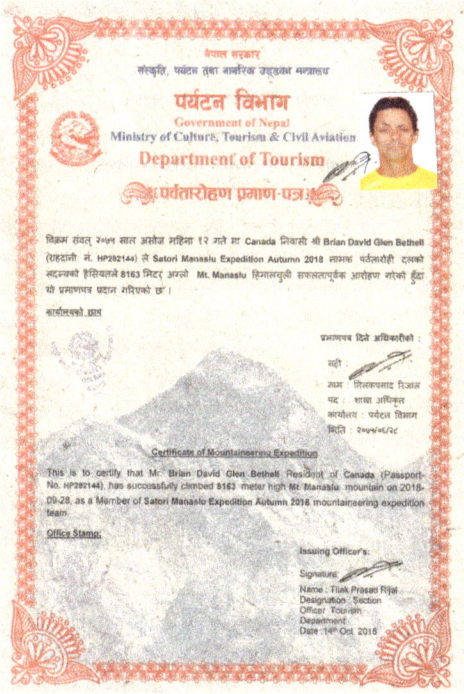

My climb is official and in the record books.

Almost two years after creating the Canadian Rockies Youth Society, I'm so happy to be able to finally fly its flag from Manaslu's summit. It's absolutely amazing to be looking down upon the Himalayan peaks surrounding us; they seem to be on their knees, bowing to us with great reverence and respect.

I'm feeling no issues with the altitude at all, and it's a stunning beautiful day to reach the summit. I'm so proud that all my sacrifice and hard work has come to fruition. I can also sense the presence of my departed mom and dad's spirits and their pride within me. This makes me so very happy. So happy I could burst, or maybe I am just swelling from the altitude.

We go back down to the fore summit. Then talk with other climbers, take pictures, and prepare, as we are only at the halfway point. The most dangerous part is ahead of us: most people have accidents and die of exhaustion or altitude illness on the way down.

Going down is much more abusive to the body. The impact of constantly down-stepping is harder on your joints and ligaments, aggravated further by exhaustion and weakened muscles. You also have significant weight on your back as you collect all your gear from the four camps on the way down. The only benefit to descending is that the air becomes thicker with more available oxygen.

We make pretty good time getting to Camp Four, stopping briefly to rest and collect our supplies, including the tent we stayed in. As we continue, I take the oxygen mask off as the air is much thicker now. At Camp Three, we talk with Pema Sherpa as he is dismantling his and Taro's camp. I'm out of water, so Pema fires up his stove and melts some snow for all of us. We hear that a Czech climber is missing near Camp Four; sherpas are searching but haven't found him yet.

Taro is still in the tent, methodically packing his gear as he doesn't want to leave anything behind. I see that Pema has not changed from last year: he is still in a hurry, trying to push Taro into hurrying up. Well, Taro, isn't having any part of this, and starts to push back from inside the tent. I start laughing. Pema spins and glares at me, which makes me laugh even harder. Then I hear Taro laughing inside the tent. Priceless!

Pasdawa and I leave for Camp Two to sleep for the night. It seems to take us no time at all—the lower we get, the more energy we regain from the increase in oxygen levels.

We arrive at Camp Two and already many tents are missing. Some people are actually going all the way down to basecamp from the summit. Good on them, but we are not in any hurry. I'm not here to only get the summit and leave safely. For me, this is a journey as well as an experience, and again I want to enjoy and savour it.

We immediately crawl into our tents and go to sleep for the night.

Saturday, September 29, 2018

I wake after a great sleep, eat a quick breakfast, then pack my things. There's an update on the missing Czech climber. We now hear that it was two climbers missing, and they have been found dead inside their tent at Camp Four.

I think they must have perished from altitude illness, mixed with exhaustion, the cold, and maybe inadequate ventilation in the tent, and as the carbon dioxide built up from them exhaling would just reduce the oxygen levels even more. Apparently, they were climbing without sherpa support, which would have made all the difference—a sherpa would have known what to watch out for. Hypoxia is deceptive, because you essentially become drunk, to the point that you have no clue where you are or that you are about to die.

On our way down to Upper Camp One, we descend through the icefall. The blocks of ice crack, pop and shift as they warm up in the blazing sun. It's not long before we arrive at Upper Camp One and again, many tents are already gone. A group of sherpas are taking tents down and dismantling the camp, packing it all away to be hauled down to basecamp on their backs. Pasdawa stops and chats with his sherpa brothers while I rest, drink more water, and eat a couple granola bars.

Kami Rita Sherpa who holds the world record of twenty-four summits on Everest is also here with his client Vanessa who says she made the summit and we congratulate each other. Pasdawa and I break up camp, and I now have all the gear that I brought up from basecamp on my back, including two sleeping bags. my total load weighs about thirty-five kilos.

I see the German nurse who told me to turn around because of my cough. She congratulates me when I tell her I made the summit, and that I'm feeling much better now. I ask how it went for her. She tells me that just before Camp Four, her blood oxygen levels dropped to 38%, which is dangerously low. She came back down here to Upper Camp One to recover for the night, and says that she's going back up tomorrow. I don't have the heart to tell her that it's supposed to be windy and cold. I wish her luck, saying the mountain will always be there for her if she doesn't make it this year. She says to me that she has to make it this year. To me this statement hints of summit fever. I hope she stays safe.

I leave for basecamp ahead of Pasdawa as he is still visiting his friends. Once I get off the steep section below Lower Camp One and onto the crevasse field of Manaslu's glacier, the sun plus its reflection off the snow is blazing hot. I stop to reapply sunscreen and pull out the umbrella, which works magnificently, and I'm on my way hopping over crevasses. As I'm essentially alone right now, I make sure to stay clipped to the rope, just in case as some of the crevasses have opened even wider with the heat of the sun. Some take a pretty good leap to get over!

Taro is ahead of me, and I catch up with him when he stops to rest. We have a good chuckle over Pema trying to hurry him up yesterday. He tells me that Pema has been pushing him all through the expedition. He finally had enough and got into it with Pema, telling him, "I pay your wages, and you must stay with my pace, not the other way around!"

I laugh and say the same thing happened between Pema and me last year—he would go about two hundred feet ahead of me and just stand there and watch me, or even go so far ahead that he was completely out of sight.

Taro asks me about Pasdawa, and I say that he's awesome: very patient and helpful, not to mention sacrificing himself for my well-being when my regulator froze. Not once did he say anything about how slow I was, especially when all his sherpa friends were already at the camps while he was stuck waiting for me.

Then Pasdawa passes us—and he is *moving*! Taro can hardly believe how much gear Pasdawa is packing on his back.

"Holy crap, is he ever strong!" says Taro.

"I think he has three tents and all the supplies from two camps on his back," I say. It looks like at least fifty or sixty kilos. Taro and I watch in awe as Pasdawa continues on down the glacier.

I get back on my feet and leave Taro to enjoy the sun. I'm getting very warm and am tired of holding the umbrella, so I put it away. I take my jacket off, but have long sleeves on and kept those down.

To stay cool, I scoop up handfuls of snow and squeeze them as I walk. I alternate hands as I'm still using my trekking pole, and the melting snow helps a lot against the heat. I cant believe how warm it was as the sweat was just rolling off of me.

I keep forgetting that I'm still at an altitude with lower oxygen levels. A few times, I get moving too fast, and need to slow down so my lungs can catch up.

Pasdawa and I finally arrive back at the crampon point. We take off the crampons and harnesses, then sit in the sun and drink a couple of Cokes stashed inside the plastic gear barrel. Coke never tasted better, although it's Indian Coke, with a different flavour than our Coke in North America as it has a different kind of sweetness to it. Still hot, I keep grabbing snow and squeezing it in my hand to cool myself. Gelzen comes up from basecamp with some juice and snacks for us.

After this, we head back to basecamp, and Gelzen helps to carry some of the gear. Once in basecamp, my batteries shut right down, so to speak—I'm completely exhausted. It's difficult to muster the strength to just drag the pen across the paper of my journal.

François has been here since yesterday, as he and Radu came right down to basecamp from the summit. He tells me Radu had already left basecamp for Kathmandu in a helicopter. François, Taro, and I will be trekking most of the way back to Arughat with our sherpas, then take a jeep the rest of the way to Kathmandu. But first we will stay in basecamp for the night, go back to Samagaun tomorrow, then start for Kathmandu the following day. While in Samagaun, I will take the toys to the Sama School for the children.

Porters have already left for Samagaun with our gear bags. Mingma Tenzi says our gear bags will be travelling back to Kathmandu through Dharipani and Besisahar. The same route we trekked to here last year. This is a longer route, so our gear bags might arrive in Kathmandu a couple days after us.

Dinner is a chicken sizzler that was quite good. For dessert, the cook makes a cake with icing that says "Congratulations for Manaslu success Sept 28, 2018." The cake was very good and sweet which immediately gave me a boost, Francois and I sat and talked for a while.

Before long, it's dark out and time to go to bed for my last night on this the eighth highest peak in the world. Locked in slopes and ice of this mountain, as well as my soul, are many memories! As I lie in my sleeping bag,

I hear lots of celebrating, partying, and loud dance music coming from the Chinese camp. It sounds like they are having a blast; every once in a while, I can hear one of them howling like a wolf.

Sunday, September 30, 2018

I had a great sleep last night, and get up to my usual routine of firing up the heater in the common tent before breakfast and having a coffee. After breakfast, I start craving Coke again. I ask our cook, but we're out of Coke, and also juice.

I go down to my buddy Lakpa's Pioneer Adventures camp, and ask him if they had a spare Coke. Even though he's busy organizing the teardown of their camp and distributing the gear to the porters, he takes a minute to chat with me. Lakpa says that there are landslides the way we will be trekking out. This may be why our heavy bags are going the other way—it'll be easier and safer for the porters on this route. Lakpa has fourteen team members who made it to the summit this year, and they are all going down to Samagaun this afternoon. Samagaun will be very busy for a couple days.

Back at our camp, I write a couple entries in my journal before getting ready to leave for Samagaun. I have a good look around, as I will probably not come back up here again. I am going to miss this place. I think of all the memories I have here in basecamp and on Manaslu, the Mountain of the Soul. Thank you for teaching me so much about myself, and goodbye to you, my friend.

As we leave basecamp, clouds roll in. We are in the clouds for ten minutes before they pass us by. Back in the sunshine, we pass a few porters as we descend. Once we get back to the treeline, I stay right behind Mingma Tenzi. We are moving fast and making great time, as it only takes us about two hours to reach the edge of the village.

CHAPTER 21:
THIS IS BECOMING A BAD HABIT

Once we get into Samagaun, we go straight to the Tashi Delek Hotel. I see that my gear bags are here, along with the bag of toys. Lakpa and his group with Pioneer Adventures are also here, and it's good to see them. I get the same soundproof room that I had previously. I put my backpack in the room, then I go down to the dining room to join Taro and François.

As we're talking, I feel some tingling in my fingers. I glance at my hands, and some of my fingers are discoloured and swollen.

Holy crap, it looks like I have frostbite! Both forefingers and both middle fingers are affected. The worst is my left-hand middle finger, which is black up to the second knuckle and feeling numb. The middle finger on my right hand is black to the first knuckle, dark blue to the second knuckle, and semi-numb. Both forefingers are dark blue to the first knuckle and tingling.

How did my fingers become frostbitten? There were no signs of it at basecamp, and it was warm outside the rest of the way down. Unless it happened up on the mountain and hasn't come to light until now? Not once did my fingers freeze or even feel cold. Even while we were waiting so long on the summit.

I begin to fear the worst, amputation! I show Mingma Tenzi and François, who both have much experience with treating frostbite. Mingma says I should soak my fingers in a bowl of lukewarm salt water. He gets the bowl of water for me and adds the salt, telling me to protect my fingers and avoid hitting them against anything. François tells me to take some aspirin to thin my blood and help increase blood flow.

I take the aspirin and soak my fingers for about an hour as we talk. The colour begins to return to my forefingers, but my middle fingers aren't improving much. We conclude that I should not trek out; I need to get to a hospital and have my fingers treated as soon as possible.

I laugh nervously and say, "I will be leaving here again in a helicopter and going back to the hospital. Oh my God, this is becoming a bad habit!"

Mingma Tenzi calls Lakpa at the office and sets up a flight to Kathmandu for tomorrow. I tell Mingma that I still need to get the toys to the children at the Sama School. He says we should do this now, as I will need to be at the helipad early tomorrow morning. He and Pasdawa will come with me to carry the bag of toys, as I need to be extremely careful with my fingers.

Mingma asks the hotelkeeper for directions to the school and learns its outside of the village, about a twenty-minute walk. When we flew in over the school, it didn't seem that far, but distances can look different when you're flying.

We head out. I feel bad that the guys must carry the bag for me, but at least it isn't heavy. About fifteen minutes later, we pass some grazing yaks and see the school.

We enter through the stone gates into the large bi level courtyard, where about a hundred children aged between four and fourteen are playing. As we walk in with the big red gear bag, the children spot us and quiet down, curious about our arrival. We walk to a small building that has its door open and the guys set the bag down on the ground in the courtyard. Mingma talks with the gentleman inside, who appears to be cooking rice. Good thing Mingma and Pasdawa are with me, as this man doesn't speak English and I don't speak Sherpa.

Mingma tells the man why we're here and he seems pleased. Some of the kids have caught the conversation, and word has spread: excited children are gathering around the red bag as it sits on the ground.

Our only dilemma is that there are fewer stuffed toys in the bag than children. I feel terrible that I don't have a toy for each child. Mingma suggests we give the toys to the youngest children, aged four to six, who would receive the most comfort from them. This works out, as I have more than enough toys for this age group.

The children of the Sama school with the toys and me.

The man and Mingma tell the younger children to circle the bag. I kneel down beside the bag and look up to see I am encircled by all of these chattering children. I open the bag, tear the plastic inner bag open, and start to hand the toys out to the little ones. The children are so happy—I can hardly hear my own thoughts, with their excited talk entering my ears and going straight to my heart. It's the most wonderful chatter I've ever heard in all my life.

I feel the presence of my mom and dad smiling down upon me, as they were generous and compassionate towards children themselves. I can only hope that my parents know all their hard work and sacrifice to raise me wasn't for naught, and they can be proud of me now and as I continue to help children who are suffering.

It may have taken me a long time, but I now know that the most powerful force in the universe is compassion given unconditionally! Giving the toys to the children for me was so much better than standing on Manaslu's summit, as this touched my heart and soul to my core like nothing else in my life has ever done before.

We walk back to the village with huge smiles on our faces. Once back at the hotel, I soak my fingers in warm saltwater again as we all talk for a while more.

Off to bed now, *subha ratri*!

Monday, October 2, 2018

I had a great sleep last night, but wake up this morning anxious about my fingers. I slept with my big mitts on my hands to protect my fingers and keep them warm. I take off the mitts and see they haven't improved much and neither has the numbness. This really worries me!

I say a little prayer then get my things ready to fly. All of my gear will be flying with me in the Kathmandu express, a Heli Everest AS350 B3e helicopter.

I eat breakfast with François and Taro, then say my goodbyes as they leave to begin their trek out. I wish I could go with them; I will have to come back and finish this other half of the trek.

Mingma Tenzi comes to the helipad, but can't wait with me as he needs to catch up to the others trekking out. I hang out with the daughter of the hotel's owner, who's waiting for some Tuborg beer for the hotel to come from Kathmandu. My buddy Lakpa of Pioneer Adventures and his group of sherpas are also here, as they have gear being flown down from basecamp. There are also locals at the helipad waiting for supplies from Kathmandu, and a couple local men with five-gallon jugs of jet fuel to refuel the helicopter.

Mingma Tenzi with me at the helipad.

The helicopter flies up from Kathmandu and lands. Lakpa and a couple of his guys offload more beer and supplies for the hotels. The helicopter needs to make some trips between Samagaun and basecamp before it will go back to Kathmandu with me on board. After about six cycles to basecamp and back, which takes over three hours, it's my turn. They refuel the bird and Lakpa places my bags in the back. I tell Lakpa I'll be in touch, then get into the co-pilot seat, strap myself in, and close my door.

Kathmandu express. My ride back to Kathmandu to have my frostbitten fingers treated

Leaving Samagaun.

The pilot spools up the engine. After we lift off, he points the nose down and we gain speed. We fly forward between a couple big trees and over the roofs of the stone houses. I'm feeling solemn, as I love this village and its souls. But I know that I'll be coming back—not to climb Mt Manaslu again, but to bring more toys and goodies for the children, and maybe help the rest of the village. Namaste to you, Samagaun and dhan'yavad!

Man! Once again, nothing compares to flying through the deep, narrow gorges that slice between these mighty yet majestic peaks. The mountains ascend into the heavens above, while their snow melts and cascades in spectacular waterfalls, plummeting hundreds of feet to the valleys below. Their cool spray dances and sparkles in the warm Himalayan sunshine forming holographic rainbows.

Before long, we are flying over Kathmandu on our final approach into the airport. The helipad here is as busy as usual, and the outside temperature gauge says it's 28°C. Not nearly as hot as when we landed here last year, but then again it's later in autumn and therefore a little cooler.

Pemba Thendu Sherpa is here to meet me, as his brother Lakpa Sherpa the managing director of Adventure 14 Peaks is busy. Getting out of the helicopter doesn't feel nearly as bad as last year—this time I'm not suffering from heatstroke or altitude illness! For the second time, I'm able to experience the fastest way through Kathmandu traffic, in the back of an ambulance with the siren blaring. I'm starting to feel like a celebrity here.

In the ambulance, concerned about the outcome of my fingers.

The ambulance brings us to the CIWEC Hospital, then Pemba and I go into emergency admitting. I only wait five minutes and the nurses take me right in, unlike at home where you wait for four to eight hours.

The doctor comes in to see me and asks me about my fingers. She has me hold my hands out, palms up, with my eyes closed. She drags her fingers lightly over mine and asks if I can feel it. I tell her yes, but my left middle finger has much less feeling.

She examines this finger more closely. She says I have a bad case of frostbite, but it doesn't seem serious enough to warrant anything more than protecting it and keeping my fingers clean and dry. What a relief!

She also says that taking the aspirin and soaking my hands in warm salt water more than likely saved my fingertips, and asks me how I knew to do this. I tell her about Dr. François's advice to take the aspirin, and Mingma preparing the warm saltwater for me.

She explains that sometimes frostbite can take time to show itself, so it could have happened sometime earlier when I was on the mountain. At higher altitudes with lower oxygen, fingers do not need to completely freeze for the tissue to start dying. Then the cause of my frostbite dawns on me. I think it was when I was squeezing the snow in my hands to cool myself as I came from down the glacier! But I don't mention this to the doctor as I feel kind of embarrassed thinking it was self induced.

She tells me to take care, and asks that I come back to see her if it does not improve in a week. She says the nurses will clean and bandage my finger, and give me some supplies; I should change the bandage daily, applying pure aloe vera.

The two nurses come in, soak my finger in Betadine, then put a big tube bandage on it. I go back to the waiting room to join Pemba. He's happy to hear that everything should be okay, and calls for Nabin, the driver of the hotel, to come and get us. (Yes, there are two Nabins at the hotel! One is the manager, Nabin Giri, and the other is Nabin Chhetri the driver.)

The traffic is bad at this time of day. The hospital is only fifteen minutes away from the hotel, but it takes Nabin just over an hour to arrive. Back at the hotel, I'm greeted with big smiles and warm hugs, as the Giri family I feel is now my family. I go to my room and have the most amazing hot shower of my life, albeit with a surgical glove over my left hand.

When I see myself in the mirror, I almost run away scared. I've lost so much weight on the mountain with the illness! Between eating poorly and my body beginning to consume itself from the exertion of climbing at altitude, I look like Skeletor.

I laugh after I put my jeans on, because I can easily pull them down over my hips even after they're buttoned up. Thankfully, I have a suit to wear to Sarmila's mother's house, and it has a belt. I belt my jeans to ensure they won't fall down as I walk down the street, which would surely create mass hilarity for everyone in view, especially if they dropped and I tripped on them. I imagine this would be even funnier than my ripped jeans back in Jasper.

After my shower, I go upstairs to my favourite rooftop café. I bask in the warm sunshine and the sweet scent of flowers on the breeze, eating chicken momos that are simply out of this world. Yummy, yummy, yummy!

After dinner, I spend some time checking my messages. I've received a tidal wave of congratulations from all over the world, and people are touched by my compassion for the children of Samagaun.

I now get the official story about the Czech climbers who died in Camp Four. In fact, it was just one Czech climber; he's missing and still has not been found. They think he fell on the exposed traverse just below Camp Four. If that's the case, he would have taken the grand tour, so he is presumed dead. Godspeed to our brother, and prayers for his loved ones.

Time for bed now. *Subha ratri.*

Tuesday, October 3, 2018

I had a great sleep. It's so nice to be on a real bed with nice soft pillows! I wake up just before 5:00 am, then fall back asleep until 7:00 am. I then get up, shower, get dressed and head upstairs. It felt so good to shave all this growth off my face.

I eat my breakfast on the roof. All the little birds hopping around, flitting about and chirping, reminds me of a Disney movie. Taro shows up at the hotel, as he's finished the trek and back from the mountain, and while at the hotel we usually have breakfast together and swap stories about the mountain.

With Sarmila at the Garden of Dreams.

Leonardo Proverbio, who was going to ski down from the summit, ended up only skiing from Camp Four. He didn't make it to the summit because of the high winds on the 29th. He said that only a few people made it to the summit that day so I am thinking the German nurse may not have realized her dream on the summit.

I spend a few more days in Kathmandu with Sarmila as my guide. We visit a few places in the city such as the zoo, which is still rebuilding from the earthquake. We also visited the Garden of Dreams and the Chandragiri Hills cable cars up to a Shiva Temple on the top of a mountain that overlooks Kathmandu. After a few days, she and I hop on a microbus and go to Panchkal, where she was raised.

With Ama Giri in Panckal during the Dashain festival, who adopted me as one of her own and most wonderful soul I have ever met.

Panchkal is a farming community, about two hours away by bus, over the hills and into another wide lush valley. I stay for two weeks with Sarmilas brother Kumar and his family, Samjhana his wife, Samrina and Sabnam his daughters who adopt me as one of their own. They are so gracious in sharing their home, food, and Dashain festival with me. Dhan'yavad to all of you.

I am forever grateful, and deeply love all of my family in Nepal—especially Ama Giri, Sarmila and Nabin Giri's mother. I spent many hours with Ama sitting in the sunshine and watching the goats, birds, and butterflies.

I cannot wait to come back.

That's all for now until Mt Everest

Namaste

CONCLUSION

This is the end of my first book. First and foremost, I want to thank you, my reader, for your support. My hope is that you feel you made a good choice in purchasing this book. I also hope you found these stories entertaining, as I went to great lengths to record my daily memoirs.

I am so happy that I was able to go back to Mt Manaslu to complete the climb, especially after what happened during my first attempt. Arriving pre-acclimatized made all the difference, and I believe I can be successful on Mt Everest. At the same time, the moments that brought the most joy to my soul was bringing toys to the children of Samagaun. Because of all the wonderful adventures and memories Manaslu has brought me, it has a special place in my heart, and has also given me the ability to share these stories of Elton, my life, my mission, and my journeys with you.

Of course, I cannot forget that Manaslu was my gateway to Everest. Without climbing Manaslu, attempting Mt Everest wouldn't be a wise idea. Knowing that I feel comfortable and healthy above 8,000 metres, I can make a real go of climbing Everest. With this I will be seeking funding and sponsorship to help with this and help disadvantaged children and youth through Canadian Rockies Youth Society, in memory of my friend Elton. This book is part of the fundraising effort, it is also meant to raise awareness for these children/youth.

I'm now in the next phase of my mission: training and fundraising for Mt Everest. Climbing the big Himalayan peaks is all about legs, lungs, and acclimatization. Marathoners usually do well at high altitudes, so I'm running two to three days a week, for at least one hour each day. In addition to regular weight training, I will soon begin extended high-altitude training. About a week out from leaving for Mt Everest, I will begin writing down my daily memoirs again so that I can write a follow-up to this book. I am sure that Everest will bring about many more stories to entertain you.

My mission is to create awareness of the struggles for disadvantaged inner-city children and youth, in memory of Elton. I also want to get a grave marker made honouring his short life that inspired me to help other kids. My original plan was to send disadvantaged inner-city kids to the Centre for Outdoor Education, based in Nordegg, Alberta, but this is a large complicated task with many moving parts. For now, I will have to work up to this ultimate goal. I'm new to all of this; I have much to learn and a lot of work ahead of me. Through the sales of this book and other fundraising avenues, I will donate to charitable children/youth organizations.

I will also continue taking toys and learning supplies to the children of Samagaun in Nepal. I also need to help the people of Samagaun, once it's feasible for me to do so. I'm researching a few ideas, like having a local person employed to build and install small stoves and chimneys in the unheated houses, or raising money to build a medical centre in this remote area. I also hope to use the sales of this book to offset my cost of climbing Mt Everest (about $40,000 USD), although I plan to raise the majority of this myself.

If, God willing, I do very well in fundraising, then I would like to set up the Canadian Rockies Youth Society as a full society or foundation in Elton's name. The goal of this Society would be to get inner-city kids to the mountains—out of the city and away from their stressors and social media—to teach them back-country skills, and just let them be kids where they can feel safe.

Finally, I want thank you again for your support. I am so sincerely grateful to you; it means everything and I cannot do this without your help. I'll see you again in my next book, about Mt Everest.

Dhan'yavad mero sathi (thank you, my friend)

GLOSSARY

A

Abalakov: The original name for V-thread ice anchor, named after its innovator, Soviet climber Vitaly Abalakov

acclimatize: To become accustomed to a new climate or new conditions, particularly your body's adjustment to less available oxygen.

alveoli: Parts of the lungs responsible for introducing oxygen to the bloodstream.

altitude illness/acute mountain sickness (AMS): Negative health effect of high altitude, caused by acute exposure to low amounts of oxygen at high altitude.

approaching: Hiking or trekking to a climbing route.

ascender: A rope grab with a handle that only slides up the rope, not down. Used to aid in climbing fixed ropes. Also called a jumar.

autorotation: A state of flight in which the main rotor system of a helicopter or similar aircraft turns by the action of air moving through the rotor. Used when the engine fails and the helicopter freefalls until close to the ground. The pilot then increases pitch of the rotor blades and uses the built-up inertia in the rotor system to set the helicopter down for a soft, safe landing.

B

basecamp: A camp from which mountaineering expeditions set out.

belay: Fundamental technique of using a rope to arrest a climber's fall, using a belay device that applies friction to the rope.

bergschrund: A large, gaping crevasse at the top of a glacier. Usually caused by an angle change in the slope, forcing the glacier to break away from the snow cap as the slope becomes steeper.

bivouac/bivy: French, meaning "temporary encampment."

C

cairn: A pile of rocks used as a route marker.

carabiner: Shackle with a spring-loaded gate used to clip to ropes, anchors, and climbing gear.

Cessna 152: American two-seat, fixed tricycle gear, general aviation airplane, used primarily for flight training and personal use.

cirque: A half-open, semicircular, steep-sided hollow at the head of a valley or on a mountainside, formed by glacial erosion.

col: The lowest point on a mountain ridge between two peaks.

coldstore: A store with refrigeration to keep ice, drinks, and items cold, used primarily in warmer climates.

cornice: An overhanging edge of snow on a ridge or crest of a mountain.

crampons: Metal plates with spikes that fix to climbing boots for use on ice or alpine rock climbing.

crevasse: A chasm that splits a glacier and can be thousands of feet deep.

D

Dashain: One of the most important Hindu festivals, celebrated all over Nepal. Usually falls in the month of September or October and is celebrated for fifteen days.

death zone: Found on mountains above 8,000 metres, a zone where there is insufficient oxygen to support life.

drainages: Gullies down the sides of mountains that have been eroded by water flow.

drytooling: Climbing on rock using ice axes and crampons.

E

exposure: Indicates the risk of injury in the event of a fall, based on the steepness of the terrain. High exposure is more dangerous.

G

griffon: The Himalayan vulture in the family Accipitridae. Closely related to the European griffon vulture and a true raptor.

grand tour: Falling high on the mountain that results in death near the base of the mountain due to being unclipped to the rope.

H

hypoxia: Condition in which the body, or a region of the body, is deprived of oxygen at the tissue level. The brain's ability to reason becomes impaired. Results in death if not alleviated.

I

ice axe/tool: An axe used by climbers to chop steps in snow and ice and as a handhold when sharp end is swung and stuck in the ice.

icefall: A steep part of a glacier that looks like a frozen waterfall. Usually is the most dangerous area of a mountain, with big seracs and house-size blocks of ice constantly shifting and moving downslope.

ice screw: A tubular, hollow screw cored into ice to provide an anchor when climbing.

J

jumar: See "ascender."

K

Kama sutra: Ancient Indian Sanskrit text on sexuality, eroticism and emotional fulfillment.

katabatic breeze/wind: A wind that carries high-density air down a slope of a mountain. Cool dense air descending from a glacier to the warmer valley below.

M

mixed climbing: similar to drytooling, with the addition of ice to climb on as well as rock. See "drytooling."

N

no-thread: Similar to a V-thread anchor, but used primarily for rappelling. Involves running the end of a rappelling rope through the ice until it reaches into two equal lengths. See "V-thread."

P

packrat: Another name for a woodrat.

picket/snow picket: A sharpened T-shaft about a metre long driven into the snow as an anchor or to attach fixed climbing ropes to.

prominence: The difference in elevation between the peak of a mountain and its lowest point in the surrounding terrain or valley floor.

protection: Devices or items used in anchor systems to catch a climber should they take a fall.

puja: an expression of worship or devotion, primarily associated with Hinduism and Buddhism.

R

rappel: To descend steep or vertical terrain by sliding down a rope, using a belay device to control your speed by applying friction to the rope.

S

scrambling: Unroped, off-trail travel that requires some use of hands.

scree: Loose slope of rock fragments, irregularly shaped and about the size of golf balls.

serac: A tower of ice on a glacier or mountain, or a block of ice in the icefall of a glacier that can shift without warning.

Sherpa: An ethnic group of Himalayan people living on the borders of Nepal and Tibet, who migrated from the high plateaus of Tibet to Nepal, renowned for their strength, skill in mountaineering and their ability to deal with very high altitudes. Also known for their warm smiles and their beautiful happy souls. When capitalized, "Sherpa" refers to a name or ethnic group; when uncapitalized, "sherpa" is a more general term for a professional Himalayan mountain guide.

Shiva: The third god in the Hindu triumvirate, which consists of three gods responsible for the creation, upkeep, and destruction of the world.

snow bridge: A bridge of snow over a crevasse, usually formed by the wind.

summit fever: The compulsion to reach the summit of a mountain at all costs.

T

talus: Baseball-sized, irregularly shaped rocks that form slopes due to erosion of the mountain.

traditional climbing/trad: A style of rock or alpine climbing where the lead climber places gear in the rock for fall protection, and the following climber collects this gear while being belayed by the lead climber above.

traverse: A lateral movement when climbing or descending, going mainly sideways rather than up or down.

V

V-thread: An anchor formed in ice by using a 22-cm ice screw to bore two conjoining holes in the ice that form a V shape. A 1/2 metre piece of rope is passed through the hole and tied together to form a loop through the ice then used as an anchor

CPSIA information can be obtained
at www.ICGtesting.com
Printed in the USA
BVHW022154080620
580680BV00005B/3